Demand System Specification and Estimation

Demand System Specification and Estimation

ROBERT A. POLLAK

TERENCE J. WALES

New York   Oxford
OXFORD UNIVERSITY PRESS
1992

Oxford University Press

Oxford   New York   Toronto
Delhi   Bombay   Calcutta   Madras   Karachi
Kuala Lumpur   Singapore   Hong Kong   Tokyo
Nairobi   Dar es Salaam   Cape Town
Melbourne   Auckland
and associated companies in
Berlin   Ibadan

Published by Oxford University Press, Inc.,
200 Madison Avenue, New York, New York 10016

Oxford is a registered trademark of Oxford University Press

Library of Congress Cataloging-in-Publication Data
Pollak, Robert A., 1938–
Demand system specification and estimation / Robert A. Pollak, Terence J. Wales.
p.   cm.   Includes bibliographical references and index.
ISBN 0-19-506941-2
1. Demand functions (Economic theory)
2. Consumer behavior—Mathematical models.
I. Wales, Terence J.   II. Title.
HB820.P65   1992   91-16054
338.5'212—dc20

9 8 7 6 5 4 3 2 1

Typeset by Thomson Press (India) Ltd., New Delhi, India

Printed in the United States of America
on acid-free paper

To Vivian
and
To Wendy

Preface

In this book we explore some of the important issues involved in bridging the gap between the pure theory of consumer behavior and its empirical implementation. The theoretical starting point is the familiar static, one-period, utility maximizing model in which the consumer allocates a fixed budget among competing categories of goods. The associated demand system is derived from the direct or indirect utility function and its parameters estimated on the basis of a set of price–quantity observations. Many issues must be addressed in moving from the theoretical model of utility maximization to estimation and interpretation of demand system parameters. We have not attempted a comprehensive treatment of all of the issues that arise in empirical demand analysis, but instead have focused on four issues that we believe are of primary importance: the assumptions made about the functional form of the utility function (i.e., the structure of preferences), the treatment of demographics, the treatment of dynamics, and the specification of the stochastic structure of the demand system. These four issues have implications for the empirical implementation of the theory of consumer behavior and for the interpretation of estimated demand systems.

In Chapter 1 we use a simple demand system—the linear expenditure system (LES)—to introduce the terminology and notation we use throughout the book. We use the LES to illustrate the four issues that we discuss in detail in four following chapters.

In Chapter 2 we discuss classes of functional forms for utility functions and for demand systems that play important roles in empirical demand analysis. Our main objectives are to elucidate the relationships among particular functional forms and to show how particular forms can be constructed from a set of basic building blocks. We begin by discussing demand systems classified by the role of expenditure. In addition to expenditure proportionality (i.e., homothetic preferences), we consider demand systems that are linear and quadratic in expenditure, and share systems that are linear and quadratic in the log of expenditure. We then discuss separability (e.g., direct additivity, indirect additivity, weak separability). We conclude Chapter 2 by discussing flexible functional forms (e.g., the translog, AIDS).

In Chapter 3 we describe five general procedures for incorporating demographic variables into demand systems. Each of the procedures replaces an original demand system by a related class involving additional parameters and postulates that only these additional parameters depend on the demographic variables. These procedures can be used to model the effects on consumption patterns not only of demographic variables but also of any other variables (e.g., environmental variables such as air quality) that affect demand patterns. We conclude Chapter 3 by discussing the role of demographic variables in welfare analysis, and, in particular, the problem of comparing the welfare of households with different demographic profiles.

In Chapter 4 we analyze two dynamic models based on changing tastes. We first discuss habit formation, examining alternative procedures for incorporating habits into arbitrary demand systems and studying their implications for short-run and long-run demand behavior. We then discuss interdependent preferences, a dynamic specification in which preferences and demand depend on the consumption patterns of other individuals or households. Finally, we consider the implications of these dynamic demand specifications for welfare analysis.

In Chapter 5 we discuss the stochastic specifications that form the basis of the empirical analysis described in Chapters 6 and 7. We present our standard stochastic specification for demand systems in share form—additive, independent (across observations but not goods), normal errors with 0 mean and a constant nondiagonal contemporaneous covariance matrix—and extend it in several directions. These extensions include first order vector autoregressive systems (we pay particular attention to the treatment of the first observation), error component models in which disturbances have a time-specific component as well as a general component, and random coefficients models. Our objective in Chapter 5 is not to provide a comprehensive survey of stochastic specifications but to lay the groundwork for our two empirical chapters.

In Chapter 6 we report demand system estimates based on a series of annual household budget data sets. Our results shed light on some of the specification issues discussed in the first five chapters. In particular, we compare demand system functional forms, and various procedures for incorporating the number of children and an indicator of their age distribution. We also explore several dynamic and stochastic specifications, including a simple error components model with a time-specific effect, and a random coefficient model for two different demand systems.

In Chapter 7 we report estimates based on annual aggregate time series data. As with the household budget data, we compare various functional forms and dynamic specifications. We also investigate estimation procedures that differ in their treatment of the first observation. Finally, we explore the possibility of pooling data from different countries under various assumptions about the existence of short-run and/or long-run differences among countries. We estimate a pooled model using data from Belgium, the U.K., and the U.S. and find that the data reject pooling.

We have ignored many interesting and important issues relevant to empirical demand analysis. On the theoretical side, for example, we have ignored recent advances in the theory of aggregation and in the modeling of intertemporal demand, both with and without separability. On the estimation side, we have ignored recent work on nonparametric and semi-parametric methods that have been proposed as alternatives to the parametric approach we have followed in this book. We have also ignored recent work on the analysis of qualitative or discrete choice. These important issues deserve careful treatment in a treatise on empirical demand analysis, but we have not attempted to write such a treatise. Instead, we have examined carefully four basic issues that must be dealt with in empirical implementation of the pure theory of consumer behavior. The reader, having grasped the treatment of these basic issues, will be well-positioned to explore other issues.

In writing this book we have accumulated debts to many people and are grateful for the advice and comments of our colleagues. We thank in particular John Bigelow, Angus Deaton, Howard Howe, Dale Jorgenson, Lawrence Lau, Jan Magnus, Michael McCarthy, David Ryan, and Robert Summers. We are also grateful to the National Science Foundation and the Social Science and Humanities Research Council of Canada for supporting our research and to Judith Goff for exceptional editorial assistance.

Seattle   R. A. P.
Vancouver   T. J. W.
May 1991

Contents

Demand System Specification and Estimation

1

Introduction

We begin this chapter with an extended example based on the linear expenditure system (LES). We use the LES to introduce the four aspects of demand system specification that we discuss in detail in the next four chapters: functional form specification (Chapter 2), the role of demographic variables (Chapter 3), dynamic structure (Chapter 4), and stochastic structure (Chapter 5). In Chapters 6 and 7 we present estimates based on these specifications using household budget data and per capita time series data.

1. FUNCTIONAL FORM: THE LES

In this section we introduce the LES and establish the terminology and notation we use throughout the book. Consider a direct utility function, $U(X)$, of the form

$$(1) \quad U(X) = \sum_{k=1}^{n} a_k \log(x_k - b_k), \quad a_i > 0, \quad (x_i - b_i) > 0, \quad \sum_{k=1}^{n} a_k = 1,$$

where x_i denotes the quantity of good i and n the number of goods. It is useful to think of the goods as broad aggregates such as food and clothing, rather than as narrowly specified commodities such as bread, butter, and jam, although the theory does not require this or any other interpretation. We shall use the terms "goods" and "commodities" interchangeably.[1]

We write the budget constraint as

$$(2) \qquad\qquad \sum p_k x_k = \mu.$$

---

[1] Distinguishing between goods and commodities is important when discussing household production. Becker's [1965] household production model postulates that households combine "market goods" such as food or clothing (bread or running shoes) with time to produce "basic commodities" such as health and prestige; these commodities are the arguments of the household's utility function. Becker's terminology is now firmly established and, although ordinary English usage suggests reversing Becker's use of "goods" and "commodities," it is now too late. Michael and Becker [1973] provide a sympathetic restatement of the household production model; Pollak and Wachter [1975] emphasize its limitations.

When writing summation signs, we shall often omit the index and limits of summation, adopting the conventions that the omitted index is k and that the summation runs from 1 to n (i.e., over all goods).

Maximizing the utility function (1) subject to the budget constraint (2) yields the ordinary (i.e., Marshallian) demand functions, $h^i(P, \mu)$,

$$(3) \qquad x_i = h^i(P, \mu) = b_i - \frac{a_i}{p_i} \sum p_k b_k + \frac{a_i}{p_i} \mu$$

where the p's denote prices and μ denotes total expenditure on the n goods, hereafter called "expenditure." At first glance the LES appears to have 2n parameters (n b's and n a's) but, because the a's must satisfy the normalization rule $\sum a_k = 1$, only $2n - 1$ of the LES parameters are independent. We shall often write demand functions in "expenditure form"

$$(4) \qquad p_i x_i = p_i b_i + a_i \left(\mu - \sum p_k b_k \right).$$

Indeed, the LES owes its name to the fact that expenditure on each good is a linear function of all prices and expenditure.

As our use of the term "expenditure" in place of "income" suggests, we interpret demand theory as a model of expenditure allocation among an exhaustive set of consumption categories. Our terminology, although not yet standard, is convenient in empirical demand analysis because it enables us to distinguish between a household's receipts in a particular period (i.e., its "income," as reported in household budget data) and its total spending on the consumption categories included in the analysis (i.e., its "expenditure" in our terminology). Somewhat inconsistently, we defer to well-established usage and refer to "income–consumption" curves rather than "expenditure–consumption" curves.

A demand system consistent with utility maximization is said to be "theoretically plausible." The LES is the only theoretically plausible demand system for which expenditure on each good is a linear function of all prices and expenditure. This result was established by Klein and Rubin [1947–1948] in their paper introducing the LES. Samuelson [1947–1948] subsequently showed that the LES could be derived from the direct utility function (1). Because the type of argument used to establish the Klein–Rubin theorem can be applied to other demand system specifications, we prove this result in Appendix A.

The budget share devoted to good i, which we denote by w_i, is obtained by dividing the expenditure form of the demand equation by μ; we denote the budget share equations by $\omega^i(P, \mu)$. For the LES the budget share equations are given by

$$(5) \qquad w_i = \omega^i(P, \mu) = \frac{p_i b_i}{\mu} + a_i \left[1 - \frac{\sum p_k b_k}{\mu} \right].$$

The LES is transparent in the sense that its parameters have straightforward behavioral interpretations. A household whose demand system is an LES is often described as first purchasing "necessary," "subsistence," or "committed" quantities of each good $(b_1, .., b_n)$, and then dividing its remaining or "supernumerary" expenditure, $\mu - \sum p_k b_k$, among the goods in fixed proportions $(a_1, \ldots, a_n)$. Thus, the quantities $(b_1, \ldots, b_n)$ can be interpreted as a "necessary basket." In any demand system the "marginal budget shares" are defined as the fractions of an additional dollar of expenditure spent on each good:

(6)
$$\frac{\partial p_i h^i(P, \mu)}{\partial \mu}.$$

Marginal budget shares must sum to 1 and, for noninferior goods, are nonnegative. For the LES the marginal budget shares are constants, that is, they are independent of prices and expenditure, and they are equal to the a's. Goldberger [1969] has proposed a useful characterization of the LES in terms of marginal budget shares: the LES is the only demand system generated by an additive direct utility function that exhibits constant marginal budget shares.

Own-price, cross-price, and expenditure elasticities of demand for the LES are easily calculated. Let $E_j^i(P, \mu)$ denote the elasticity of demand for good i with respect to p_j and $E_\mu^i(P, \mu)$ the expenditure elasticity:

(7)
$$E_i^i(P, \mu) = \frac{p_i b_i (1 - a_i)}{p_i b_i + a_i \left(\mu - \sum p_k b_k \right)} - 1$$

(8)
$$E_j^i(P, \mu) = \frac{- a_i b_j p_j}{p_i b_i + a_i \left(\mu - \sum p_k b_k \right)}$$

(9)
$$E_\mu^i(P, \mu) = \frac{a_i \mu}{p_i b_i + a_i \left(\mu - \sum p_k b_k \right)}.$$

Because the LES price and expenditure elasticities are functions of all prices and expenditure rather than constants, these elasticities do not provide a transparent summary of the behavior implied by a particular set of LES parameters. Indeed, because the LES parameters have a straightforward behavioral interpretation, the parameter values themselves provide the most transparent summary statistics for the LES.

What does the LES indifference map look like? We begin with the Cobb–Douglas, which is a very simple special case of the LES. The Cobb–Douglas direct utility function can be written as

(10a)
$$U(X) = \sum a_k \log x_k, \qquad a_i > 0, \qquad \sum a_k = 1,$$

or, equivalently, in its more familiar "constant returns to scale" form

(10b) $$V(X) = \prod x_k^{a_k}, \qquad a_i > 0, \qquad \sum a_k = 1.$$

Figure 1 illustrates the indifference map corresponding to the Cobb–Douglas. The corresponding ordinary demand functions are given by

(11) $$x_i = h^i(P, \mu) = \frac{a_i}{p_i}\mu,$$

or, in expenditure form,

(12) $$p_i x_i = p_i h^i(P, \mu) = a_i \mu.$$

The Cobb–Douglas demand functions imply that consumption of each good is proportional to expenditure or, equivalently, that the income–consumption curves are rays from the origin. In the Cobb–Douglas case, the marginal budget shares are constants and are equal to the average budget shares.

The indifference map corresponding to the LES utility function, (1), is a Cobb–Douglas indifference map with the origin "translated" to the point $(b_1, \ldots, b_n)$. This indifference map is shown in Figure 2.

The indifference map in Figure 2 helps clarify two problems we have thus far ignored. The first, the "limited-domain problem," arises because the LES utility function is defined only in the region of the commodity space northeast of $(b_1, \ldots, b_n)$; the income–consumption curves are straight

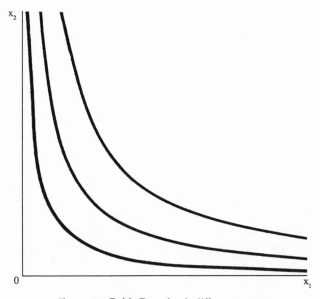

Figure 1   Cobb-Douglas indifference map

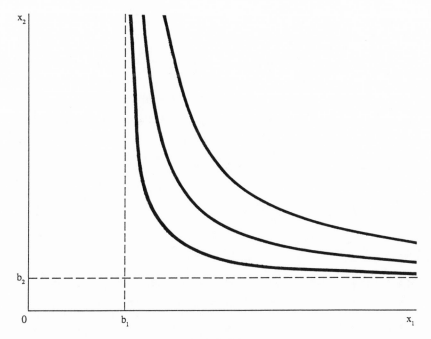

Figure 2 LES indifference map

lines radiating upward from this point. If the household can purchase a commodity bundle in this region, then it will do so; if it cannot, then, because the utility function is not defined if $x_i \leqslant b_i$ for any i, we can say nothing about the household's behavior.

If we try to finesse the limited-domain problem by beginning with the demand functions rather than with the utility function, then the limited-domain problem manifests itself in a different form — as a violation of "regularity conditions." The Slutsky matrix is defined as the n × n matrix of substitution terms $S(P, \mu) = [s^{ij}(P, \mu)]$, where

$$(13) \qquad s^{ij}(P, \mu) = \frac{\partial h^i(P, \mu)}{\partial p_j} + h^j(P, \mu) \frac{\partial h^i(P, \mu)}{\partial \mu}.$$

A preference ordering is said to be "well-behaved" if it is representable by a direct utility function that is monotonically increasing in all its arguments, has continuous second partial derivatives, and is strictly quasi-concave.[2] A demand system derivable from a well-behaved preference ordering satisfies three regularity conditions. First, the budget

---

[2] The requirement that the utility function be strictly quasi-concave, $U(X^a) \geqq U(X^b)$, $X^a \neq X^b$, implies $U[\lambda X^a + (1 - \lambda)X^b] > U(X^b)$ for all λ, $0 < \lambda < 1$, corresponds to the requirement that the indifference curves have the "right" curvature and no linear segments.

constraint holds as an identity

(14) $$\sum p_k h^k(P, \mu) \equiv \mu.$$

Second, the demand system is homogeneous of degree 0 in prices and expenditure

(15) $$h^i(\lambda P, \lambda \mu) = h^i(P, \mu), \qquad \text{for all } \lambda > 0.$$

Third, the Slutsky matrix, $S(P, \mu)$, is symmetric and negative semidefinite. Furthermore, any demand system satisfying these regularity conditions is theoretically plausible, that is, it can be rationalized by a well-behaved preference ordering. For the LES the substitution terms are given by

(16) $$s^{ii}(P, \mu) = -\frac{a_i}{p_i^2}(1 - a_i)\left(\mu - \sum p_k b_k\right)$$

(17) $$s^{ij}(P, \mu) = \frac{a_i a_j}{p_i p_j}\left(\mu - \sum p_k b_k\right), \qquad i \neq j.$$

At price–expenditure situations for which μ is less than $\sum p_k b_k$, the LES own-substitution terms are positive, a violation of regularity conditions.

The second problem that the indifference curve diagram helps clarify, the "nonnegativity problem," arises because the conventional LES demand functions, (3), were derived without regard for nonnegativity constraints on consumption; hence, in some price–expenditure situations, the conventional LES demand functions predict negative consumption of some goods. Of course, this is nonsense: the conventional demand functions are correct only when they coincide with the true LES demand functions (i.e., those derived taking proper account of the nonnegativity conditions) and this occurs at and only at those price–expenditure situations for which the conventional demand functions predict non-negative consumption of all goods. The case in which all of the b's are negative provides the clearest example; with two goods, the point (b_1, b_2) lies in the third quadrant. The income–consumption curves in the first quadrant are linear, and their linear extensions pass through the point (b_1, b_2); thus, even when the b's are negative we can describe the income–consumption curves as radiating upward from the point $(b_1, \ldots, b_n)$. When the b's are negative, however, we must abandon our heuristic interpretation of these parameters as necessary, subsistence, or committed quantities.

To find the true demand functions corresponding to price–expenditure situations for which the conventional LES demand functions predict negative consumption of some goods, we must determine which goods will not be consumed, drop them from the analysis, and use the suitably reduced LES to calculate the demand for the remaining goods. Both theoretical and empirical analysis are simplified if we ignore the nonnegativity problem and restrict our attention to a region of the price–expenditure space in which only the goods in some prespecified subset are consumed in strictly positive quantities. To achieve this simplification, we appeal to the friendly fairy who helps economists: she tells us at the outset which goods to drop from the analysis. In empirical work nonnegativity constraints can be ignored when using data for broad commodity groups that have been aggregated over households. At the other extreme, we must confront the nonnegativity problem when using household level data for narrowly defined goods.

When all the b's are positive, the conventional LES demand functions predict positive consumption at price–expenditure situations for which the regularity conditions hold. Hence, when all the b's are positive, the limited-domain problem takes precedence over the nonnegativity problem. When all the b's are negative, the LES utility function defines a well-behaved preference ordering over the entire commodity space, so the limited-domain problem does not arise. In this case, however, there must exist a region of the price–expenditure space in which the nonnegativity problem arises.

All the b's need not be of the same sign. Examination of the elasticity formulas (7), (8), and (9), confirms that demand for the ith good is inelastic if b_i is positive and elastic if b_i is negative. When the x's are broad commodity groups, we would expect inelastic demand and, hence, positive b's. The signs of the b's are an empirical question, however, and cannot be determined on the basis of a priori speculation.

The indirect utility function provides a representation of preferences that permits a straightforward derivation of the ordinary demand functions using Roy's identity. An indirect utility function represents a preference ordering over the price–expenditure space in the same way that a direct utility function represents a preference ordering over the commodity space. The indirect utility function, $\psi(P, \mu)$, can be obtained by substituting the ordinary demand functions into the direct utility function:

$$(18a) \qquad \psi(P, \mu) = \max_{\sum p_k x_k \leq \mu} U(X) = U[h(P, \mu)].$$

It is often convenient to define "normalized" prices, $y_i = p_i/\mu$, and to write the indirect utility function in normalized price form as

$$(18b) \qquad \phi(Y) = U[h(Y, 1)].$$

An indirect utility function (written in terms of normalized prices) corresponds to a well-behaved preference ordering if it is monotonically decreasing in all its arguments, has continuous second partials, and is strictly quasi-convex.[3] A major advantage of the indirect utility function

---

[3] A function $\phi(Z)$ is said to be strictly quasi-convex if $-\phi(Z)$ is strictly quasi-concave. Thus, if the function $\phi(Y)$ is a well-behaved indirect utility function, then the function $-\phi(X)$ is a well-behaved direct utility function. Except in very special cases, these two utility functions correspond to different preference orderings; we return to this issue in Chapter 2.

is that closed-form expressions for the ordinary demand functions can be obtained from it using Roy's identity:

$$(19a) \qquad\qquad h^i(P, \mu) = -\frac{\partial \psi(P, \mu)/\partial p_i}{\partial \psi(P, \mu)/\partial \mu}.$$

In terms of normalized prices, Roy's identity becomes

$$(19b) \qquad\qquad h^i(Y, 1) = \frac{\partial \phi(Y)/\partial y_i}{\sum y_k [\partial \phi(Y)/\partial y_k]}.$$

The indirect utility function corresponding to the LES is a good illustration. It is easily verified that applying Roy's identity to the indirect utility function

$$(20) \qquad\qquad \psi(P, \mu) = \frac{\mu - \sum p_k b_k}{\prod p_k^{a_k}}$$

yields the LES ordinary demand functions (3). Hence, (20) is the indirect utility function corresponding to the LES. To obtain this indirect utility function by substituting the ordinary demand functions into the direct utility function, it is convenient to begin by subjecting the LES direct utility function (1) to the increasing transformation $V(X) = T[U(X)]$, where $T(z) = e^z$. The original utility function $U(X)$ and the transformed utility function

$$(21) \qquad V(X) = \prod(x_k - b_k)^{a_k}, \qquad a_i > 0, \qquad (x_i - b_i) > 0, \qquad \sum a_k = 1$$

are equally valid representations of the same underlying preference ordering. Substituting the ordinary demand functions (3) into (21) yields

$$(22) \qquad\qquad V[h(P, \mu)] = \frac{\prod a_k^{a_k}}{\prod p_k^{a_k}} \left(\mu - \sum p_k b_k \right).$$

Dropping the irrelevant constant factor, we obtain the LES indirect utility function (20). It is unnecessary to transform the direct utility function in this or any other way in order to obtain the indirect utility function, but doing so yields a form of the indirect utility function compatible with forms that appear in Chapter 2.

As we shall see repeatedly in Chapter 2, functional form specification for a demand system usually begins with an assumption about the form of the indirect utility function and uses Roy's identity to derive the ordinary demand functions. Similarly, theorems that characterize the class of preferences compatible with some class of ordinary demand functions (e.g., demand functions linear in expenditure) usually characterize preferences in terms of the indirect utility function.

We consider now very briefly some results obtained from estimating the LES using U.K. budget study data. The sources and definitions of the variables are discussed in detail in Chapter 6. Briefly we have three broad

Table 1 Own-Price Elasticities: Static Model without Demographic Effects

S. per week	Food	Clothing	Miscellaneous
200	−.66	−1.41	−1.53
300	−.73	−1.23	−1.26
400	−.78	−1.15	−1.17

Notes:
1. These elasticities are calculated on the basis of LES parameters estimated from U.K. budget study data (81 observations).
2. Own-price elasticities are evaluated at 1970 prices.
3. Mean expenditure in 1970 is 315 S. per week.

consumption categories—food, clothing, and miscellaneous—for the years 1968–1972. These data are cross-classified by income level and family size, yielding a total of 81 observations. For the moment we ignore the information about family size and attempt to explain the observed differences in consumption patterns using only prices and total expenditure as explanatory variables. We obtain a stochastic form for the LES by adding a disturbance term to each of the share equations given in (5). We denote the 3×1 vector of disturbances corresponding to the ith cell by $u_i = (u_{i1}, u_{i2}, u_{i3})'$ and assume that $E(u_i) = 0$, that $E(u_i u_j') = \Omega$ for all i and j, and that the u_i are independently normally distributed.

In Table 1 we present the own-price elasticities obtained from estimating the LES with the specification just described using the maximum likelihood procedure. The demand for food is inelastic at all three expenditure levels, while demand for the other two goods is elastic. Further, the demand for food becomes less inelastic as expenditure rises, while the demands for clothing and for miscellaneous become less elastic. Although not shown in Table 1, the (constant) marginal budget shares for these three goods are .33, .22, and .45, respectively, while the values of b are 66.4, − 14.4, and 41.4 shillings (S.) per week.

2. DEMOGRAPHIC SPECIFICATION

Demographic variables such as family size and age composition have traditionally played a major role in the analysis of household budget data. Family size and composition, race, religion, age, and education have all been used as demographic variables in demand studies, although only recently in the context of complete demand systems.

There are two ways to allow for demographic variables. First, given enough data, it is always possible to estimate separately the demand systems for subsamples of households with identical demographic profiles. For example, we might specify that each household's demand equations are given by an LES and estimate the $2n - 1$ independent parameters of that system separately for each household type. This approach allows all of the parameters of the demand system to depend on the demographic profile and does not require us to specify the form of the relationship between the parameters and the demographic variables. Under this approach the only data relevant to the analysis of households with a particular demographic profile are observations on households with that profile.

The second approach, which we discuss in Chapter 3, introduces specifications that relate the behavior of households with different demographic profiles. Here we describe only one such specification, "demographic translating," which we introduce in the context of a single demographic variable—for definiteness, household size, which we denote by η. We assume that each household's demand equations are given by an LES, that the b's depend linearly on the number of persons in the household,

$$(23) \qquad\qquad b_i = b_i^* + \beta_i \eta,$$

and that the a's are independent of the demographic variables. The implied demand system (in expenditure form) is given by

$$(24) \qquad x_i = h^i(P, \mu, \eta) = b_i^* - \frac{a_i}{p_i} \sum p_k b_k^* + \frac{a_i}{p_i} \mu + \beta_i \eta - \frac{a_i}{p_i} \sum p_k \beta_k \eta.$$

To examine the effect of household size on the consumption of good i we differentiate (24) with respect to η, obtaining

$$(25) \qquad\qquad \frac{\partial h^i(P, \mu, \eta)}{\partial \eta} = \beta_i - \frac{a_i}{p_i} \sum p_k \beta_k.$$

The budget constraint implies that a change in household size that causes an increase in the consumption of one good must also cause offsetting decreases in the consumption of other goods. A major advantage of working with complete demand systems rather than analyzing the demand for each good separately is that the complete system approach restricts our attention to reallocations of expenditure that satisfy the budget constraint. The effect of an increase in family size on the consumption of a particular good must thus be viewed in the context of allocating expenditure among all goods. In terms of our LES example, even the direction of the effect of an increase in household size on consumption of the ith good cannot be inferred from the direction of the effect of such a change on the demand system parameter b_i. Indeed, there is no presumption that an increase in household size will increase rather than decrease b_i, since changes in the b's, regardless of their direction, imply a reallocation of expenditure among the goods but leave total expenditure unchanged.

To illustrate the translating procedure, we used the data set consisting

Table 2 Own-Price Elasticities: Static Model with Demographic Effects

Family size/ S. per week	Food	Clothing	Miscellaneous
One child			
200	−.73	−1.86	−1.83
300	−.78	−1.48	−1.42
400	−.81	−1.33	−1.28
Two children			
200	−.66	−1.83	−1.79
300	−.72	−1.45	−1.41
400	−.76	−1.31	−1.28
Three children			
200	−.60	−1.79	−2.23
300	−.67	−1.41	−1.49
400	−.71	−1.28	−1.30

Notes:
1. See Notes to Table 1.

of 81 observations introduced in Section 1 to estimate Eq. (24) in share form. We now use the cross-classification by family size as well as by income. In Table 2 we present the estimated own-price elasticities for different expenditure levels and family sizes. The pattern of results, for a given family size, is the same as that obtained without using family size as an explanatory variable: the demand for food is inelastic and becomes less inelastic as expenditure rises, while the demand for clothing and for miscellaneous become less elastic. A comparison of elasticities across family sizes indicates that the number of children has very little effect, although in all but two cases demand becomes slightly less elastic as family size increases, for given expenditure levels. Although not reported in the table, the marginal budget shares for the three goods are .29, .23, and .48, respectively. These are very close to the values obtained earlier with no demographics included, namely .33, .22, and .45.

The advantage of an approach that relates the behavior of households with different demographic profiles is that it enables us to draw inferences about the behavior of households with one demographic profile from observations on the behavior of households with different profiles. Inferences of this type are not possible if we estimate separately the demand systems for households with each type of demographic profile. This advantage is especially important when degrees of freedom are limited, either because the number of observations is small or because the number of demographic variables is large. In Chapter 3 we examine demographic translating in more detail and show that it is a "general procedure" in the sense that it can be used in conjunction with any complete demand system. We also examine other general procedures for incorporating demographic variables into complete demand systems and discuss the problem of comparing the welfare of households with different demo-graphic profiles.

3. DYNAMIC SPECIFICATION

Empirical demand analysis must either assume that demand system parameters remain constant over time or specify how they change. If one takes the necessary basket interpretation seriously, then it seems plausible that the b's might vary over time. From a technical standpoint, it is relatively simple to incorporate changing b's into the LES because they enter the demand functions linearly.

Although the b's in the LES can sometimes be interpreted as necessary quantities, there is no presumption that they are physiologically rather than psychologically necessary. Indeed, it seems plausible that the "necessary" quantity of a good should depend, at least in part, on past consumption of that good. The simplest assumption is that the necessary quantity of each good is proportional to consumption of that good in the previous period, that is,

$$(26) \qquad\qquad b_{it} = \beta_i x_{it-1}$$

where b_{it} is the value of b_i in period t and β_i a "habit formation coefficient." A more general assumption is that the necessary quantity of each good is a linear function of consumption of that good in the previous period, that is,

$$(27) \qquad\qquad b_{it} = b_i^* + \beta_i x_{it-1}.$$

Here b_i^* can be interpreted as the "physiologically necessary" component of b_{it} and $\beta_i x_{it-1}$ as the "psychologically necessary" component. This specification provides a simple example of a "habit formation" model.

If all goods are subject to habit formation of the type described by (27), then the utility function becomes

$$(28) \qquad U(X_t; X_{t-1}) = \sum a_k \log(x_{kt} - b_k^* - \beta_k x_{kt-1}).$$

The semicolon separating X_t and X_{t-1} indicates that the preference ordering over current consumption (X_t) is "conditional" on past consumption (X_{t-1}); we discuss the interpretation of this habit formation specification in Chapter 4. In period t the individual is supposed to choose $(x_{1t}, \ldots, x_{nt})$ to maximize (28) subject to the budget constraint

$$(29) \qquad\qquad \sum p_{kt} x_{kt} = \mu_t.$$

The resulting demand system (suppressing the time subscripts on the p's and μ) is of the form

$$(30) \qquad h^i(P, \mu; X_{t-1}) = b_i^* - \frac{a_i}{p_i}\sum p_k b_k^* + \frac{a_i}{p_i}\mu + \beta_i x_{it-1} - \frac{a_i}{p_i}\sum p_k \beta_k x_{kt-1}.$$

These short-run demand functions, like their static counterparts (3), are locally linear in expenditure. Because the b's are linear in past consumption and because current consumption depends linearly on the b's, present consumption of each good is a linear function of past consumption of all goods. Furthermore, provided the β's are positive, an increase in past consumption of a good implies an increase in present consumption of that good,

$$(31) \qquad \frac{\partial h^{it}(P, \mu; X_{t-1})}{\partial x_{it-1}} = \beta_i - a_i\beta_i > 0,$$

and a decrease in present consumption of every other good,

$$(32) \qquad \frac{\partial h^{jt}(P, \mu; X_{t-1})}{\partial x_{it-1}} = - a_j p_i \beta_i/p_j < 0.$$

We define "long-run" or "steady-state" demand functions corresponding to these short-run demand functions. For every price–expenditure situation, the short-run demand functions define a mapping of the consumption vector in period t into the consumption vector in period $t + 1$. A long-run equilibrium or steady-state consumption vector corresponding to a particular price–expenditure situation is a consumption vector that, if it prevailed in period t, will prevail in period $t + 1$. In Chapter 4 we show that if the short-run demand system is an LES, then the long-run demand system is also an LES, although the parameters of the long-run LES depend on the habit coefficients as well as on the other parameters of the short-run LES. We also show that these dynamic demand functions are locally stable provided the β's are all less than 1. In addition, Chapter 4 discusses alternative specifications of habit formation for the LES and for general demand systems.

An alternative dynamic specification, interdependent preferences, postulates that a household's preferences depend on the consumption of some (and perhaps all) other households. Again the LES provides a simple example. Specifications in which the demand system parameters depend on the lagged rather than the current consumption of others are more tractable and seem to capture the essential features of interdependent preferences; hence, in our discussion of interdependent preferences, we emphasize lagged interdependence. In Chapter 4 we discuss models of interdependent preferences, examining both their short-run and long-run implications for the LES and for more general demand systems.

To illustrate the dependence of the b's on some measure of past consumption, we use the U.K. data employed earlier and estimate equation (30) in share form. For the lagged quantities in dynamic demand equation (30) we use a weighted average over all observations in the sample in the previous year, because those data do not report average lagged consumption for each observation.[4] In Table 3 we present estimated own-price

---

[4] Because the observations are cell means, we weight each observation by cell size.

Table 3 Own-Price Elasticities: Dynamic Model

S. per week 2-child family	Food	Clothing	Miscellaneous
231	−1.13	−2.66	−2.66
259	−1.12	−2.41	−2.35
280	−1.11	−2.26	−2.19
320	−1.10	−2.06	−1.96
360	−1.09	−1.92	−1.82
476	−1.08	−1.65	−1.56

Notes
1. See Notes to Table 1.

elasticities for 1970 for families with two children for expenditure levels prevailing in that year. The elasticities are all higher in absolute value than those given in Tables 1 and 2—indeed the demand for food is now slightly elastic. The marginal budget shares, however, are identical to those in Table 1, namely .33, .22, and .45 for the three goods. The values of β, although not reported in Table 3, are .44, − .43, and .25, respectively. Thus the values for food and miscellaneous are consistent with habit formation, while the value for clothing is not.

We conclude Chapter 4 by arguing that, although endogenous preferences are tractable in both theoretical and empirical demand analysis, they cause serious difficulties for welfare analysis.

4. STOCHASTIC SPECIFICATION

Estimating a demand system requires assumptions about stochastic structure. The most common stochastic specification begins with the share form of the demand system and adds a disturbance term to each share equation. In the case of the LES this implies

$$(33) \qquad w_{it} = \frac{p_{it}b_i}{\mu_t} + a_i \left[1 - \sum \frac{p_{kt}b_k}{\mu_t} \right] + u_{it}.$$

Instead of adding disturbance terms to the share equations, one could add them to the expenditure equations. We prefer the specification based on the share form because we think it is likely to involve less heteroskedasticity.

The "standard" stochastic specification assumes that disturbances are independent across observations. In the case of time series data this implies independence over time periods; in the case of household budget data it implies independence over households.

In Chapter 5 we discuss in detail the standard specification and consider several alternative stochastic structures, including a serial correlation model, an error components model, and a random coefficients model. In a random coefficients model all observations correspond to the same parametric demand system, but some of the underlying demand system parameters are random variables. In the LES, for example, we could assume that the b's contain a random term

$$(34) \qquad\qquad\qquad b_{it} = b_i^* + v_{it}$$

so that the demand system is given by

$$(35) \qquad w_{it} = \frac{p_{it} b_i^*}{\mu_t} + a_i \left[1 - \frac{\sum p_{kt} b_k^*}{\mu_t} \right] + \frac{p_{it} v_{it}}{\mu_t} - a_i \frac{\sum p_{kt} v_{kt}}{\mu_t}.$$

We could obtain this form directly from (33) by specifying that the u's are given by

$$(36) \qquad\qquad w_{it} = \frac{p_{it} v_{it}}{\mu_t} - a_i \frac{\sum p_{kt} v_{kt}}{\mu_t}.$$

Different assumptions about the distribution of the v's will, of course, have different implications for the covariance matrix of the u's. In Chapter 5 we discuss a number of such assumptions and their implications. In Chapter 6 we use the U.K. data to estimate this random coefficients specification under various assumptions about the distribution of the v's. We do not present the results here because, under what we think are the most plausible assumptions about the distribution of the v's, the estimated own-price elasticities and marginal budget shares are virtually identical to those reported in Table 1.

5. ESTIMATION

Data never speak for themselves. A unifying theme of this book is that empirical analysis depends on both the data and the specification of a theoretical model. In empirical demand analysis the selection of a theoretical model entails choosing a functional form, a demographic specification, a dynamic specification, and a stochastic specification. The necessity of choosing a functional form is inescapable, although proponents of flexible functional forms have sometimes suggested that, by choosing such a form, one can avoid imposing a priori restrictions on admissible demand behavior. Choosing a demographic specification can be avoided only by estimating separate demand systems for households with distinct demographic profiles. Choosing a dynamic specification cannot be avoided: the assumption that demand system parameters remain constant over time, which is the only alternative to modeling explicitly how they change, is itself a dynamic specification. The necessity of choosing a stochastic specification has been recognized only recently; mindlessly adding a disturbance term to a nonstochastic model is no longer professionally acceptable. In summary: theory must play a crucial role in structuring empirical research and we think that this role should be explicit rather than implicit.

In Chapters 6 and 7 we turn to estimation, in Chapter 6 using household budget data and in Chapter 7 using per capita time series data. Although household budget data have long played an important role in empirical demand analysis, the analysis of such data has traditionally focused on demographic effects and expenditure or income effects, and has ignored price effects. In Pollak and Wales [1978] we demonstrated that the LES and other interesting complete demand systems can be estimated using household budget data from as few as two periods, despite the limited price variability present in such data.

Because we expect household budget data will play an increasing role in empirical demand analysis, we have emphasized two aspects of demand system specification that are especially important in dealing with such data. In Chapter 2 we emphasize functional form specifications that yield Engel curves suitable for analyzing data in which expenditure variation is substantial relative to price variation, as is typically the case in household budget data. In Chapter 3 we discuss the treatment of demographic variables, an issue that has traditionally played a central role in the analysis of household budget data.

Five conclusions emerge from our analysis of household budget data in Chapter 6.

- Although the LES is an attractive functional form because of its theoretical and empirical tractability, it is not appropriate for the analysis of household budget data. Likelihood ratio tests reject the LES restrictions in all samples against a wide range of alternative specifications.
- Functional forms with three-parameter Engel curves are significantly superior to forms with two-parameter Engel curves. The likelihood ratio test not only rejects the LES against forms allowing quadratic Engel curves, but also rejects the translog, a form involving a two-parameter Engel curve, against a three-parameter generalization.
- Demographic variables matter. We reject the pooled specification, which assumes demographic variables have no effect, against each of five procedures we describe for incorporating demographic variables into complete demand systems.
- Dynamic specifications are superior to static ones. Although this conclusion need not be interpreted as evidence of taste change, it does indicate that the static specification is flawed.
- Stochastic and dynamic specifications often act as substitutes; because they are difficult to disentangle, they should be evaluated together.

In Chapter 7 we use per capita time series data from national product accounts to estimate complete demand systems. We begin by investigating alternative specifications of functional forms, dynamic structures, and stochastic structures. We cannot easily summarize our results here because we have not yet introduced the specifications being compared. We can say, however, that there is strong evidence that dynamic factors play an important role. As with household budget data, dynamic and stochastic specifications tend to act as substitutes and, hence, it is difficult to determine the extent to which dynamic factors operate through the nonstochastic rather than the stochastic portion of the model. On balance, our results confirm the importance of dynamic factors, lending support to the emphasis we have placed on habit formation and other dynamic specifications.

We conclude Chapter 7 by exploring the possibility of pooling data from different countries for demand system estimation. The obvious objection to pooling consumption data is that we would expect two countries facing the same prices and having the same level of expenditure to have systematic differences in their consumption patterns. We allow for these systematic differences in two ways. First, we use dynamic specifications that permit "transitory" or short-run differences across countries reflecting differences in past consumption patterns. Under the transitory difference specification, if two countries were to face the same price–expenditure situation for a number of successive periods, then their consumption patterns would converge to a common value. Pooling, however, does not require that all countries have identical consumption patterns, even in the long run. Our second specification, the "permanent difference" specification, allows a subset of the demand system parameters to differ across countries and is thus consistent with underlying demographic, climatic, or taste differences. We also estimate mixed specifications that permit both transitory and permanent differences among countries.

A second difficulty in pooling data from different countries is related to their use of different national currencies. We show that when quantities are measured in different value units (e.g., U.S. food in 1970 U.S. dollars; Belgian food in 1970 B. francs) the problem is essentially the same as when goods are measured in different physical units. Pooling requires transforming the data into common units, and commodity-specific purchasing power parities are the conversion factors required to do this. Despite the attractiveness of pooling data from different countries and our choice of three countries—Belgium, the U.K., and the U.S.—for which pooling seems plausible, our empirical results do not support pooling.

6. CONCLUSION

In this chapter we have used the LES to establish the terminology and notation that appears throughout the book. In addition, the LES illustrated four basic aspects of demand specification that provide context and motivation for the extensive discussions of functional form specification, the role of demographic variables, and basic elements of dynamic and stochastic structure in Chapters 2 through 5. Finally, we used empirical estimates of the LES based on household budget data and per capita time series data to foreshadow our concern with estimation in Chapters 6 and 7 and to illustrate a major theme of this book—the crucial role of economic theory as a foundation for empirical analysis.

APPENDIX A: THE KLEIN–RUBIN THEOREM

Theorem: Let $h(P, \mu)$ be a theoretically plausible demand system in which expenditure on each good is a linear function of prices and expenditure

$$\text{(A1a)} \qquad h^i(P, \mu) = \frac{1}{p_i} \sum p_k \hat{\alpha}^{ik} + \frac{\alpha^{i0}}{p_i} \mu$$

and for which the marginal budget shares are not equal to 0 or 1 for any good. Any such demand system can be written in the LES form

$$\text{(A2)} \qquad h^i(P, \mu) = b_i - \frac{a_i}{p_i} \sum p_k b_k + \frac{a_i}{p_i} \mu, \qquad \sum a_k = 1.$$

Proof: It is convenient to rewrite the demand system (A1a) as

$$\text{(A1b)} \qquad h^i(P, \mu) = \alpha^{ii} - \frac{a_i}{p_i} \sum \alpha^{ik} p_k + a_i \frac{\mu}{p_i},$$

where the new variables are defined by

$$a_i = \alpha^{i0}$$

$$\alpha^{ij} = -\frac{\hat{\alpha}^{ij}}{\alpha^{i0}} = -\frac{\hat{\alpha}^{ij}}{a_i}$$

$$\alpha^{ii} = \frac{\hat{\alpha}^{ii}}{1 - \alpha^{i0}} = \frac{\hat{\alpha}^{ii}}{1 - a_i}.$$

Since a theoretically plausible demand system must satisfy the budget constraint, we must have $\sum a_k = 1$, as the theorem requires.

Although rewriting (A1a) and (A1b) is not crucial to proving the theorem, it does simplify the analysis. As is so often the case, it is easier to prove a characterization result when you know where you are going. Thus, we define b_t by

$$b_t = \alpha^{1t}.$$

To prove the theorem, we must show that

(A3) $$\alpha^{jt} = \alpha^{it} = b_t$$

for all j and t.

A theoretically plausible demand system must satisfy the Slutsky symmetry conditions

$$\frac{\partial h^i}{\partial p_j} + h^j \frac{\partial h^i}{\partial \mu} = \frac{\partial h^j}{\partial p_i} + h^i \frac{\partial h^j}{\partial \mu}.$$

Letting $i = 1$, for the demand system (A1b) the Slutsky symmetry conditions become

(A4) $$-\frac{a_1}{p_1} \alpha^{1j} + h^j \frac{a_1}{p_1} = -\frac{a_j}{p_j} \alpha^{j1} + h^1 \frac{a_j}{p_j}.$$

Multiplying both sides by $p_1 p_j$ yields

(A5) $$-p_j a_1 \alpha^{1j} + p_j h^j a_1 = -p_1 a_j \alpha^{j1} + p_1 h^1 a_j$$

for all j. Differentiating (A5) with respect to p_t, $t \neq 1, j$, yields

$$-a_1 a_j \alpha^{jt} = -a_j a_1 \alpha^{1t}$$

so

$$\alpha^{jt} = \alpha^{1t}$$

for all $j \neq t$. Differentiating (A5) with respect to p_j yields

$$-a_1 a_j \alpha^{j1} = -a_j \alpha^{j1} + a_j \alpha^{11} - a_j a_1 \alpha^{11},$$

or, equivalently,

$$(a_j - a_1 a_j)\alpha^{j1} = (a_j - a_j a_1)\alpha^{11}$$

so

$$\alpha^{j1} = \alpha^{11}.$$

This establishes (A3).   QED

2

Functional Form Specification

Functional form specification is a central aspect of empirical demand analysis.[1] Our primary objective in this chapter is to elucidate the relationships among particular parametric forms by showing that particular functional forms can be viewed as members of families or classes of functional forms. Our secondary objective is to demonstrate how particular forms can be constructed from a set of basic building blocks.

We emphasize classes of functional forms rather than treating particular parametric forms in isolation for three reasons. First, a reader's guide to currently used parametric forms would be obsolete before publication. The menu of parametric forms used in empirical demand analysis has grown rapidly over the last two decades and there is every reason to expect this rapid growth to continue. Second, focusing on classes of functional forms enables the reader to place in perspective not only currently used parametric forms but also new ones as they are introduced. Critical consideration of alternative parametric forms is essential both in planning one's own research and in evaluating the work of others. Our strategy—offering an analytical taxonomy—emphasizes the relationships among particular forms and thus increases the reader's awareness of alternative specifications. Third, viewing particular parametric forms against the background of the classes to which they belong enables us to recognize more easily and economically the limitations of particular functional forms. For example, suppose we specify and estimate a parametric functional form and find that our estimates imply that food and clothing are substitutes; to interpret this result, we must know whether our parametric functional form can exhibit complementarity. If it cannot, then our finding that food and clothing are substitutes was dictated by our functional form and reflects nothing about our data or economic reality. Focusing on the behavioral implications of classes of functional forms draws attention to the behavioral restrictions they imply.

Awareness of alternative specifications will not enable us to avoid making restrictive assumptions—restrictive parametric assumptions are

---

[1] This chapter draws on Howe, Pollak, and Wales [1979], Pollak [1969, 1971a, 1971b, 1972], and Pollak and Wales [1980].

inevitable in empirical demand analysis. Such awareness, however, will enable us to recognize the behavioral restrictions inherent in particular functional forms and, hence, to select appropriate forms for particular problems. As the previous sentence suggests, we do not believe that there is a single, "one-size-fits-all" functional form that is ideal for all applications. Instead, we believe that the characteristics that make a particular functional form suitable for one application may well make it inappropriate for another. For example, household budget data typically present the investigator with wide variation in observed levels of total expenditure but limited price variation. Time series data, on the other hand, typically offer less variation in expenditure and more variation in relative prices. Thus, it is not surprising that the parametric forms best suited for analyzing household budget data differ from those best suited for analyzing per capita time series data.

Our ordering of topics in this chapter is necessarily somewhat arbitrary. In Section 1 we discuss classes of demand systems defined in terms of the role of expenditure. In addition to the homothetic case, we discuss demand systems linear and quadratic in expenditure and share systems log-linear and log-quadratic in expenditure. In Section 2 we discuss demand systems generated by preference orderings satisfying various separability assumptions. We begin with direct additivity, indirect additivity, and generalized additive separability, a class that includes direct and indirect additivity. We then discuss weak and strong separability—the separability concepts appropriate for groups of goods. In Section 3 we discuss flexible functional forms, a class that includes the generalized Leontief and all of the translog forms including the "Almost Ideal Demand System" (AIDS) proposed by Deaton and Muellbauer. Because we did not discuss flexible functional forms in Chapter 1, we begin Section 3 with a detailed examination of the translog family.

1. EXPENDITURE SPECIFICATIONS

In this section we consider classes of demand systems in which expenditure (i.e., total expenditure or income) enters in simple ways. We begin by discussing expenditure proportionality, the theoretically important but empirically improbable case in which each good's expenditure elasticity is unity. We then discuss the class of demand systems in which the demand for each good is a linear function of expenditure, an empirically useful class that includes the LES. Continuing in this vein, we consider the class of demand systems in which the demand for each good is a quadratic function of expenditure; the empirical results we report in Chapters 6 and 7 rely heavily on members of this quadratic class.

We then consider an alternative direction for generalizing expenditure proportionality. Expenditure proportionality is equivalent to the requirement that the demand system, written in share form, is independent of the level of total expenditure. Working with the demand system in share form, we generalize this by considering the class of demand systems in which shares are linear functions of expenditure raised to a power or linear functions of the log of expenditure. We also consider a further generalization of the logarithmic class in which shares are quadratic in the log of expenditure.

1.1. Expenditure Proportionality

Expenditure proportionality plays an important role in economic theory, but has little practical relevance to empirical demand analysis. "Engel's Law," one of the great empirical regularities to emerge from nineteenth-century studies of household budget data, asserts that the budget share of food is smaller for rich than for poor households. Expenditure proportionality, which implies that the budget share of every good is independent of the household's total expenditure, thus contradicts Engel's Law. Despite its simplicity, the assumption of expenditure proportionality is too unrealistic to serve even as a useful first approximation for empirical demand analysis. Hence, demand systems exhibiting expen-diture proportionality—especially the Cobb–Douglas, the Leontief, and the CES—are useful in empirical demand analysis primarily as building blocks that enter into the construction of more general demand systems.[2] The situation is different in the analysis of production where homothetic production functions, especially those exhibiting constant returns to scale, play a key role. The Cobb–Douglas, the Leontief, the CES, and the homothetic translog (which we discuss in Section 3) play key roles in the empirical analysis of production.

A demand system is said to exhibit expenditure proportionality if the demand for each good is proportional to expenditure

$$(1) \qquad\qquad h^i(P, \mu) = B^i(P)\mu,$$

or, equivalently, if all expenditure elasticities are unity. If a demand system generated by a well-behaved utility function exhibits expenditure propor-tionality, then the demand functions must be of the form

$$(2) \qquad\qquad h^i(P, \mu) = \frac{g_i(P)}{g(P)} \mu$$

where the function $g(P)$ is homogeneous of degree 1 and g_i denotes its partial derivative with respect to the ith price. We prove this result in Appendix A as part of the Gorman polar form theorem, which charac-terizes demand systems linear in expenditure. The demand functions (1)

---

[2] Expenditure proportionality plays a crucial role in the theory of aggregation over goods (see, for example, Blackorby, Primont, and Russell [1978]). Most of the standard, tractable, closed-form examples of globally defined demand systems exhibit expenditure proportionality.

are generated by a direct utility function that is homothetic to the origin, that is, a function that is an increasing transformation of a function homogeneous of degree 1:

$$(3) \qquad\qquad U(X) = T[V(X)]$$

where $T'(\cdot) > 0$ and $V(\lambda X) = \lambda V(X)$. Here, as elsewhere in the book, we sacrifice generality for tractability. In particular, we assume that demand functions and utility functions are differentiable enough to support calculus–based arguments. In characterizing increasing transformations (e.g., to define homotheticity or to investigate the class of direct or indirect utility functions representing a preference ordering) we confine ourselves to increasing transformations with strictly positive first derivatives, even though this excludes such strictly increasing transformations as $T(z) = z^3$.

The Cobb–Douglas and the CES are leading examples of demand systems exhibiting expenditure proportionality. The direct utility function for the Cobb–Douglas is given by

$$(4) \qquad U(X) = \sum a_k \log x_k, \qquad a_i > 0, \qquad \sum a_k = 1.$$

The direct utility function for the CES is

$$(5) \qquad\begin{aligned} U(X) &= -\sum a_k x_k^c, & a_i &> 0, & c &< 0 \\ U(X) &= \sum a_k x_k^c, & a_i &> 0, & 0 &< c < 1. \end{aligned}$$

The Cobb–Douglas ordinary demand functions are given by

$$(6) \qquad\qquad h^i(P, \mu) = \frac{a_i \mu}{p_i}$$

and the CES ordinary demand functions by

$$(7) \qquad\qquad h^i(P, \mu) = \frac{(p_i/a_i)^{1/(c-1)} \mu}{\sum p_k (p_k/a_k)^{1/(c-1)}}.$$

The indifference maps corresponding to these utility functions are identical to those of the Cobb–Douglas and CES production functions. The parameter c is related to the elasticity of substitution, σ, by $\sigma = 1/(1-c)$. It is easy to verify that when $c = 0$ the CES demand system reduces to the Cobb–Douglas. The Cobb–Douglas utility function is the limiting form of the CES as c approaches 0; Arrow, Chenery, Minhas, and Solow [1961], who introduced the CES production function, provide a clear discussion.

For most of the classes of demand systems we consider, direct utility function characterizations of the preferences that generate them either are unavailable or are uninformative in the sense that they fail to provide a transparent characterization of preferences. Indirect utility function characterizations, on the other hand, are widely available and play a central role in the analysis. In the case of expenditure proportionality, one can use Roy's identity to show that the demand system (2) is generated by the indirect utility function

(8)
$$\psi(P, \mu) = \frac{\mu}{g(P)}.$$

The indirect utility function corresponding to the Cobb–Douglas direct utility function (4) is given by (8) where

(9)
$$g(P) = \prod (p_k)^{a_k}.$$

The indirect utility function corresponding to the CES direct utility functions (5) is given by (8) where

(10)
$$g(P) = \left[\sum_k a_k^{-1/(c-1)} p_k^{c/(c-1)} \right]^{(c-1)/c}.$$

The limiting case of the CES direct utility function as c approaches minus infinity is the Leontief or fixed-coefficient utility function

(11)
$$U(X) = \min_k \left\{ \frac{x_k}{a_k} \right\}, \qquad a_i > 0,$$

as Arrow, Chenery, Minhas, and Solow [1961] show. The ordinary demand functions are given by

(12)
$$h^i(P, \mu) = \frac{a_i \mu}{\sum a_k p_k}.$$

Because the Leontief utility function is not differentiable, Lagrangian multipliers cannot be used to find the ordinary demand functions. A straightforward derivation relies on the fact that the ratios $\{x_i/a_i\}$ must all equal a common value. The indirect utility function corresponding to the Leontief case is given by (8) where

(13)
$$g(P) = \sum a_k p_k.$$

1.2. Demand Systems Linear in Expenditure

Consider demand systems linear in expenditure:

(14)
$$h^i(P, \mu) = C^i(P) + B^i(P)\mu.$$

Gorman [1961] has shown that any demand system that is linear in expenditure and theoretically plausible (i.e., consistent with utility maximization) must be of the form

(15)
$$h^i(P, \mu) = f_i(P) - \frac{g_i(P)}{g(P)} f(P) + \frac{g_i(P)}{g(P)} \mu$$

where f(P) and g(P) are functions homogeneous of degree 1. Because the argument can be used in a number of situations, we prove Gorman's result in Appendix A. If we require linearity to hold for all price–expenditure situations in which P and μ are strictly positive, then we are back to expenditure proportionality: the budget constraint implies $\sum p_k(f_k - fg_k/g) = 0$ and nonnegativity of consumption near 0 expenditure implies $f_i - fg_i/g \geqslant 0$. Hence, $f_i - fg_i/g = 0$ for all i. For purposes of empirical demand analysis it is usually sufficient to require linearity in a region of the price–expenditure space.

Using Roy's identity, we can easily verify that the class of demand systems, (15), is generated by an indirect utility function of the form

$$(16) \qquad \qquad \psi(P, \mu) = \frac{\mu - f(P)}{g(P)}$$

which is known as the "Gorman polar form." The difficult part of Gorman's theorem or any characterization theorem is going in the other direction—showing that any theoretically plausible demand system linear in expenditure can be written in the form (15) and finding the class of indirect utility functions that generate the entire class of demand functions.

The LES,

$$(17) \qquad \qquad h^i(P, \mu) = b_i - \frac{a_i}{p_i} \sum p_k b_k + \frac{a_i}{p_i} \mu,$$

the basis of our discussion in Chapter 1, is an example of a demand system linear in expenditure. It is generated by a Gorman polar form indirect utility function where $f(P) = \sum p_k b_k$ and $g(P) = \prod p_k^{a_k}$, $\sum a_k = 1$, so $f_i = b_i$ and $g_i/g = a_i/p_i$.

Although the class of direct utility functions yielding demand systems linear in expenditure has not been fully characterized, an important subclass consists of all direct utility functions of the form

$$(18) \qquad \qquad U(X) = T[V(X - b)]$$

where $T'(\cdot) > 0$ and $V(\lambda X) = \lambda V(X)$. A function U(X) satisfying (18) is said to be homothetic to the point $(b_1, \ldots, b_n)$. The indifference curves of U are scaled up (or scaled down) versions of a single base indifference curve and the income–consumption curves radiate from the translated origin $(b_1, \ldots, b_n)$. The Gorman polar form corresponding to a direct utility function homothetic to the point $(b_1, \ldots, b_n)$ is given by (16), where $g(\cdot)$ is the dual of the function $V(\cdot)$ and $f(P)$ is the linear function

$$(19) \qquad \qquad f(P) = \sum p_k b_k.$$

This linearity restriction on the form of the function $f(P)$ demonstrates that the class of direct utility functions homothetic to a point in the commodity space does not exhaust the class of direct preferences generating demand systems locally linear in expenditure.

The "trick" used in (18) to generate a new preference ordering from an old one by "translating" the origin from $0 = (0,\ldots,0)$ to the point $b = (b_1,\ldots, b_n)$ in the commodity space is quite general and can be applied to any direct utility function, $U(X)$. Furthermore, duality enables us to perform an analogous trick with the indirect utility function: by translating the origin in the normalized price space, we generate another new preference ordering. In Section 2 we use this trick in conjunction with indirect additivity.

Consider the members of the "homothetic to a point" subclass corresponding to the direct utility functions we discussed in conjunction with expenditure proportionality. The LES utility function

$$(20) \quad U(X) = \sum a_k \log(x_k - b_k), \qquad a_i > 0, \qquad (x_i - b_i) > 0, \qquad \sum a_k = 1,$$

is a translation of the Cobb–Douglas (4). Translating the two CES forms, (5), yields

$$(21) \quad \begin{aligned} U(X) &= -\sum a_k(x_k - b_k)^c, & a_i > 0, & \quad (x_i - b_i) > 0, & c < 0, \\ U(X) &= \sum a_k(x_k - b_k)^c, & a_i > 0, & \quad (x_i - b_i) > 0, & 0 < c < 1. \end{aligned}$$

The corresponding demand functions are of the form

$$(22) \qquad h^i(P, \mu) = b_i - \gamma^i(P)\sum p_k b_k + \gamma^i(P)\mu$$

where

$$(23) \qquad \gamma^i(P) = \frac{(p_i/a_i)^{1/(c-1)}}{\sum p_k (p_k/a_k)^{1/(c-1)}}.$$

The indirect utility functions are given by (16) where f(P) is given by (19) and g(P) by

$$(24) \qquad g(P) = \left[\sum a_k^{-1/(c-1)} p_k^{c/(c-1)} \right]^{(c-1)/c}.$$

Translating the Leontief direct utility function (11) yields

$$(25) \qquad U(X) = \min_k \left\{ \frac{x_k - b_k}{a_k} \right\}, \qquad a_i > 0.$$

The corresponding demand functions are given by

$$(26) \qquad h^i(P, \mu) = b_i - \frac{a_i}{\sum a_k p_k} \sum p_k b_k + \frac{a_i}{\sum p_k a_k} \mu.$$

These demand functions are generated by the indirect utility function (16) where f(P) is given by (19) and g(P) by (13).

Any quadratic direct utility function yields demand functions locally linear in expenditure. The additive quadratic, which goes back to Gossen (see Samuelson [1947, p. 93]), is a special case ($c = 2$) of the additive utility function

(27) $\quad U(X) = -\sum a_k(b_k - x_k)^c, \qquad a_i > 0, \qquad (b_i - x_i) > 0, \qquad c > 1,$

which yields linear demand functions of the form (22). The indifference map of (27) is homothetic to the point $(b_1, \ldots, b_n)$, which is a "bliss point," and the income–consumption curves are straight lines that converge at this point (see Figure 1). The bliss point must lie in the first quadrant or no positive x's satisfy the condition $(b_i - x_i) > 0$, which is necessary for well-behaved preferences. For the additive direct quadratic the indifference curves are concentric circles centered at the bliss point $(b_1, \ldots, b_n)$. The utility function (27) is defined only in a box-like region of the commodity space southwest of the bliss point; for values of μ greater than $\sum p_k b_k$, the own-substitution terms implied by (27) are positive, in violation of regularity conditions.

The ordinary demand functions corresponding to (27), like those corresponding to (21) are given by (22) where $\gamma(P)$ is of the form (23). We refer to (21) and (27) as the "generalized CES class," although (27) is not a generalization of a well-behaved CES utility function (e.g., if $c = 2$ and if all of the translation parameters are 0, then the implied "CES indifference map" consists of concentric circles centered at the origin; these indifference curves have the "wrong" curvature).

For demand systems linear in expenditure, the marginal budget share of each good—that is, the fraction of an extra dollar of expenditure devoted to each good—is independent of expenditure. In the special case of expendi-

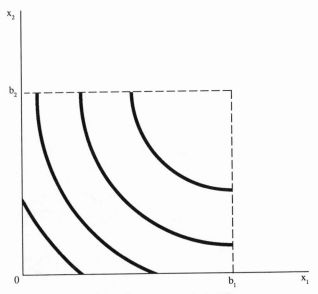

Figure 1 Additive direct quadratic indifference map

ture proportionality, marginal budget shares are not only independent of the level of expenditure but also equal to the average budget shares. For the LES the marginal budget shares are independent of prices as well as expenditure and are equal to the a's. In addition to the LES and the generalized CES, the class of demand systems linear in expenditure also includes the "linear translog" (LTL), a form we discuss in Section 3.

1.3. Demand Systems Quadratic in Expenditure

Casual empiricism suggests that the hypothesis that demand systems are linear in expenditure is too restrictive for the analysis of household budget data. Linearity implies that marginal budget shares are independent of the level of expenditure, so that rich households and poor households spend the same fraction of an extra dollar on each good. The implausibility of this hypothesis constitutes an argument for relaxing linearity so that it can be subjected to rigorous econometric testing. The class of demand systems quadratic in expenditure

$$(28) \qquad h^i(P,\mu) = A^i(P)\mu^2 + B^i(P)\mu + C^i(P)$$

provides a natural generalization. van Daal and Merkies [1989], correcting a result of Howe, Pollak, and Wales [1979], show that if a quadratic demand system is theoretically plausible, then it is of the form

$$(29a) \qquad h^i(P,\mu) = \frac{1}{\gamma}\left(\frac{g_i}{g} - \frac{\gamma_i}{\gamma}\right)\mu^2 + \left[\frac{g_i}{g} - \frac{2f}{\gamma}\left(\frac{g_i}{g} - \frac{\gamma_i}{\gamma}\right)\right]\mu$$

$$+ \frac{f^2}{\gamma}\left(\frac{g_i}{g} - \frac{\gamma_i}{\gamma}\right) - \frac{g_i}{g}f + f_i + \chi\left(\frac{g}{\gamma}\right)\frac{g^2}{\gamma}\left(\frac{g_i}{g} - \frac{\gamma_i}{\gamma}\right),$$

or, equivalently,

$$(29b) \quad h^i(P,\mu) = \frac{1}{\gamma}\left(\frac{g_i}{g} - \frac{\gamma_i}{\gamma}\right)(\mu - f)^2 + \frac{g_i}{g}(\mu - f) + f_i + \chi\left(\frac{g}{\gamma}\right)\frac{g^2}{\gamma}\left(\frac{g_i}{g} - \frac{\gamma_i}{\gamma}\right),$$

where the functions f, g, and γ are homogeneous of degree 1 and $\chi(\cdot)$ is a function of one variable. Howe, Pollak, and Wales [1979] failed in their attempt to characterize the entire class of demand systems quadratic in expenditure by implicitly assuming that $\chi(\cdot) = 0$. An alternative parameterization, instead of using the function $\gamma(P)$, uses the homogeneous of degree 1 function $\alpha(P)$ where $\alpha = g^2/\gamma$, so that the demand system (29b) becomes

$$(30) \quad h^i(P,\mu) = \frac{1}{g^2}\left(\alpha_i - \frac{g_i}{g}\alpha\right)(\mu - f)^2 + \frac{g_i}{g}(\mu - f) + f_i + \chi\left(\frac{\alpha}{g}\right)\left(\alpha_i - \frac{g_i}{g}\alpha\right).$$

Provided $\chi(\cdot) = 0$, these demand functions are generated by the indirect utility function

(31a)
$$\psi(P,\mu) = -\frac{g(P)}{\mu - f(P)} - \frac{g(P)}{\gamma(P)},$$

or, equivalently,

(31b)
$$\psi(P,\mu) = -\frac{g(P)}{\mu - f(P)} - \frac{\alpha(P)}{g(P)}.$$

The indirect utility function corresponding to the general case in which $\chi(\cdot) \neq 0$ is unknown. To facilitate comparing the indirect utility function (31) with that corresponding to demand systems linear in expenditure, we subject the Gorman polar form (16) to the increasing transformation $T^*(z) = -z^{-1}$ to obtain the equivalent indirect utility function

(32)
$$\psi(P,\mu) = -\frac{g(P)}{\mu - f(P)}.$$

Adding a function homogeneous of degree 0 to (32) yields (31).

Specifying demand systems quadratic in expenditure is one way of relaxing the assumption that demand systems are linear in expenditure. Whether a particular quadratic generalization is worth the extra cost in degrees of freedom is in part an empirical question and in part a matter of taste and judgment. Since the quadratic hypothesis nests the linear one, classical hypothesis testing provides a framework for assessing whether a particular quadratic specification represents a statistically significant improvement over the linear one. But formal tests cannot resolve whether a statistically significant improvement is worth the extra cost in terms of difficulty of estimation and forgone opportunities to generalize in other directions.

We discuss two parametric specifications of demand systems quadratic in expenditure, the λ-QES and the Σ-QES. For both of these specifications $\chi(\cdot) = 0$ and the functions $f(P)$ and $g(P)$ are given by

(33)
$$f(P) = \sum p_k b_k$$

and

(34)
$$g(P) = \prod p_k^{a_k}, \qquad \sum a_k = 1.$$

For the λ-QES, the function α is given by

(35)
$$\alpha(P) = \lambda g(P)^2 / \prod p_k^{c_k}, \qquad \sum c_k = 1;$$

for the Σ-QES, the function α is given by

(36)
$$\alpha(P) = \sum p_k c_k.$$

These specifications, like any QES specification for which $\chi(\cdot) = 0$ and f and g are given by (33) and (34), reduce to the LES when $\alpha = 0$, or, equivalently, when $\gamma(P)$ is proportional to $g(P)$.

1.4. Polynomial, Logarithmic, and Fractional Demand Systems

We open this section by discussing Muellbauer's characterization of two-term polynomial demand systems. We then introduce a general procedure for introducing constant terms (i.e., terms independent of expenditure) into theoretically plausible demand systems. We apply this procedure to Muellbauer's two-term polynomial demand system, obtaining polynomials with additional terms. We then discuss a remarkable theorem of Gorman that demonstrates the inherent limitations of polynomial demand systems. We conclude by discussing logarithmic and fractional demand systems.

The two-term polynomial demand system

$$(37) \qquad\qquad h^i(P, \mu) = D^i(P)\mu^\sigma + B^i(P)\mu$$

is an attractive generalization of expenditure proportionality. Muellbauer [1975] designates this class, together with a logarithmic class that we discuss in the next section, by the acronym "PIGL" ("price independent generalized linearity").[3] He shows that theoretically plausible demand systems of the two-term polynomial form (37) are given by

$$(38) \qquad\qquad h^i(P, \mu) = \left[\frac{g_i}{g} - \frac{\gamma_i}{\gamma}\right]\frac{\mu^\sigma}{\gamma^{\sigma-1}} + \frac{g_i}{g}\mu$$

where the functions $g(P)$ and $\gamma(P)$ are homogeneous of degree 1 and shows that these demand systems must be generated by the indirect utility function

$$(39) \qquad\qquad \psi(P, \mu) = \frac{-\mu^{1-\sigma} - \gamma(P)^{1-\sigma}}{g(P)^{1-\sigma}}.^4$$

We now propose a general procedure for introducing constant terms (i.e., terms independent of expenditure) into theoretically plausible demand systems. The trick is to replace the original demand system $\{\bar{h}^i(P, \mu)\}$ by the modified demand system $\{h^i(P, \mu)\}$ where

$$(40) \qquad\qquad h^i(P, \mu) = f_i(P) + \bar{h}^i[P, \mu - f(P)]$$

and $f(P)$ is a function homogeneous of degree 1. It is easy to verify that if the original system is generated by the indirect utility function $\bar{\psi}(P, \mu)$, then the modified system is generated by the indirect utility function

$$(41) \qquad\qquad \psi(P, \mu) = \bar{\psi}[P, \mu - f(P)].$$

---

[3] When $\sigma = 0$, two-term polynomial demand systems reduce to demand systems linear in expenditure.

[4] Muellbauer [1975] is primarily motivated by a concern with aggregation over individuals. Muellbauer works with demand systems in share form and uses the expenditure function rather than the indirect utility function. He also investigates the class of two-term demand systems exhibiting "generalized linearity," $h^i(P, \mu) = D^i(P)\phi(P, \mu) + B^i(P)\mu$.

Thus, if the original system is theoretically plausible, then the new system is also theoretically plausible, at least for values of f(P) near 0.

We illustrate this procedure with two examples. First, consider an original demand system exhibiting expenditure proportionality; applying our procedure, we obtain a demand system linear in expenditure. Replacing μ by $\mu - f(P)$ in the indirect utility function, we obtain the Gorman polar form. Second, consider the quadratic members of Muellbauer's PIGL class ($\sigma = 2$). It is easy to verify that modifying the quadratic PIGL indirect utility function

$$(42) \qquad \bar{\psi}(P, \mu) = \frac{-\mu^{-1} - \gamma^{-1}}{g^{-1}} = -\frac{g}{\mu} - \frac{g}{\gamma}$$

by replacing μ by $\mu - f(P)$ yields the indirect utility function (31), a form that generates demand systems quadratic in expenditure.

It is evident that we can use this trick to generate higher degree polynomials from two-term polynomials. For example, suppose we begin with a two-term cubic demand system and replace μ by $\mu - f(P)$. The resulting system

$$(43) \qquad h^i(P, \mu) = \left[\frac{g_i}{g} - \frac{\gamma_i}{\gamma} \right] \frac{(\mu - f)^3}{\gamma^2} + \frac{g_i}{g}(\mu - f) + f_i$$

when expanded, includes cubic, quadratic, and linear terms in μ as well as a term independent of μ. Notice, however, that the quadratic terms in the resulting demand system are proportional to the cubic terms (the factor of proportionality is $-3f$). The reader can verify that if we begin with a two-term quartic, then in the modified system both the cubic and the quadratic terms are proportional to the quartic terms. Thus, the higher order polynomial terms in demand systems generated in this way from two-term systems are severely restricted.

A remarkable theorem of Gorman's [1981] implies that such restrictions are not peculiar to the particular procedure we used to generalize two-term polynomial demand systems, but are an intrinsic feature of theoretically plausible polynomial demand systems. Gorman's theorem is complex, and we shall not state it in anything approaching its full generality. We shall, however, make plausible its implication that higher order terms in theoretically plausible polynomial demand systems must satisfy severe restrictions by considering the implications of the Slutsky symmetry conditions for third degree polynomial demand systems,

$$(44) \qquad h^i(P, \mu) = D^i(P)\mu^3 + A^i(P)\mu^2 + B^i(P)\mu + C^i(P).$$

It is straightforward, although somewhat tedious, to calculate the Slutsky terms corresponding to this demand system

$$(45) \qquad \frac{\partial h^i}{\partial p_j} + h^j \frac{\partial h^i}{\partial \mu} = \frac{\partial h^j}{\partial p_i} + h^i \frac{\partial h^j}{\partial \mu}.$$

These Slutksy terms are polynomials of degree 5 in μ and, because symmetry holds as an identity in the price–expenditure space, symmetry must hold for the coefficients of like powers of μ. (To see this, differentiate repeatedly with respect to μ and note that symmetry must hold for each of these derivatives). To illustrate our point, it suffices to calculate only that portion of the Slutsky term of degree 4 in μ. (The term of degree 5 in μ is given by $3D^jD^i$, so symmetry of this term imposes no restrictions on the coefficients of the demand system). The term of degree 4 is given by

(46a) $$2A^iD^j + 3A^jD^i$$

or, equivalently,

(46b) $$[2A^iD^j + 2A^jD^i] + A^jD^i.$$

Because the term in brackets is symmetric, symmetry of (46) implies

(47a) $$A^jD^i = A^iD^j$$

or, equivalently,

(47b) $$A^j = \frac{A^i}{D^i}D^j.$$

Hence, the coefficients of the quadratic terms (the A's in (44)) must be proportional to the coefficients of the cubic terms (the D's in (44)). The reader can verify that if we begin with a quartic demand system, then the Slutsky symmetry conditions imply that both the coefficients of the cubic terms and the coefficients of the quadratic terms must be proportional to the coefficients of the quartic terms. Gorman's theorem implies that increasing the number of terms in polynomial demand systems beyond three yields surprisingly little additional generality.

The PIGL class includes a logarithmic as well as a polynomial subclass. The logarithmic subclass, which Muellbauer calls PIGLOG, can be written as

(48) $$h^i(P, \mu) = D^i(P)\mu \log \mu + B^i(P)\mu$$

or, in share form,

(49) $$\omega^i(P, \mu) = \hat{D}^i(P) \log \mu + \hat{B}^i(P).$$

Muellbauer shows that all theoretically plausible PIGLOG demand systems can be written in the form

(50) $$h^i(P, \mu) = \frac{g_i}{g}\mu - \frac{G_i}{G}[\log \mu - \log g]\mu$$

where $g(P)$ is homogeneous of degree 1, $G(P)$ is homogeneous of degree 0, and that the corresponding indirect utility function is given by

(51) $$\psi(P, \mu) = G(P)[\log \mu - \log g(P)].$$

The PIGLOG class includes not only expenditure proportionality but two functional forms that are widely used in empirical demand analysis, the log translog (log TL) and AIDS. We examine these two demand systems in detail in Section 3.

As Deaton [1981] points out, the PIGLOG class can be generalized in much the same way as Howe, Pollak, and Wales [1979] generalized the class of demand systems linear in expenditure to obtain demand systems quadratic in expenditure. Following Howe, Pollak, and Wales [1979] and van Daal and Merkies [1989], characterizing the theoretically plausible subclass should be straightforward. Because the empirical results we present in Chapter 6 indicate that the analysis of household budget data requires a demand system whose Engel curves have at least three independent parameters, we find the quadratic generalization of PIGLOG attractive.

In an important series of papers, Lewbel [1987a, 1987b, 1990] extends Gorman's theorem in several directions. Lewbel [1987a] introduces "fractional demand systems," which he defines as demand systems in which quantities or budget shares are proportional to an expression of the form: $A^i(P)\alpha(\mu) + B^i(P)\beta(\mu)$, where $\alpha(\cdot)$ and $\beta(\cdot)$ are differentiable functions of μ. The class of fractional demand systems contains demand systems linear in expenditure, some quadratic cases, PIGL, and PIGLOG. It also contains "translog-type" cases, including the "basic translog", which we discuss in Section 3. Lewbel obtains a closed-form characterization of the class of fractional demand systems and shows that, in addition to the cases just described, it contains three other cases which he calls "LOG2," "EXP," and "TAN."

Lewbel [1987b] investigates demand systems that are linear in expenditure and an arbitrary function of expenditure: $A^i(P)\alpha(\mu) + B^i(P)\mu + C^i(P)$. These "Gorman Engel Curves" are generalizations of the QES, PIGL, and PIGLOG; Lewbel establishes a closed-form characterization of the class.

Lewbel [1990] provides a closed-form characterization of "full rank demand systems." Gorman's theorem asserts that the coefficient matrix of a demand system that is linear in functions of expenditure is at most of rank three. Full rank demand systems are those with coefficient matrices of rank three. Lewbel proves that this class consists of PIGL, PIGLOG, a generalization of the quadratic, a quadratic logarithmic case, and a trigonometric case. Taken together, the Gorman and Lewbel results bind together the apparently disparate functional forms we have discussed in this section.

2. SEPARABILITY

In this section we discuss demand systems generated by separable preferences. Direct additivity and other separability assumptions, such as indirect additivity, restrict preferences by precluding certain types of specific interactions among goods. Additive preferences are most plausible when the goods correspond to broad aggregated categories (e.g., food, clothing, and recreation) rather than to narrow ones (e.g., bread, butter, and jam).

2.1. Direct Additivity

We say that a preference ordering is directly additive if it can be represented by a direct utility function of the form

$$(52) \qquad U(X) = T[\sum u^k(x_k)]$$

where $T(\cdot) > 0$. The direct utility functions corresponding to the LES, (20), and the generalized CES, (21) and (27), exhibit direct additivity.

Direct additivity, as we have defined it, is ordinal; that is, it is a property of the preference ordering, not merely a property of the particular utility function chosen to represent preferences. When preferences exhibit direct additivity, the additive "canonical form" of the direct utility function

$$(53) \qquad U(X) = \sum u^k(x_k)$$

has no better claim to being the "true" utility function than any other representation. Indeed, the phrase "true utility function" is foreign to modern demand analysis, which views utility functions as real valued representations of preference orderings—that is, as real valued functions defined on the commodity space that assign higher values to preferred commodity vectors.[5] For example, the preference ordering corresponding to the LES is equally well represented by

$$(54a) \quad U(X) = \sum a_k \log(x_k - b_k), \qquad a_i > 0, \qquad (x_i - b_i) > 0, \qquad \sum a_k = 1,$$

---

[5]In contrast to ordinal utility, "cardinal" utility does something more; for example, it may measure the intensity of an individual's preference for one basket of goods over another or compare the welfare that two individuals derive from the baskets of goods they consume. Replacing a utility function that measures intensity of preference or one that permits interpersonal comparisons by an increasing transformation of itself may alter the measure or the comparison. Hence, the set of admissible transformations of a cardinal utility function may be limited to some narrower class such as increasing linear transformations. None of this, however, is relevant to direct additivity because we have defined it as a property of the preference ordering.

or by

(54b) $\quad V(X) = \prod (x_k - b_k)^{a_k}, \qquad a_i > 0, \qquad (x_i - b_i) > 0, \qquad \sum a_k = 1,$

or by any other increasing transformation of (54a).

Provided there are at least three goods —an assumption we maintain throughout this book—a preference ordering is directly additive if and only if the marginal rates of substitution involving each pair of goods depend only on the quantities of the two goods in the pair.[6] Equivalently, a preference ordering is directly additive if and only if the marginal rates of substitution involving each pair of goods are independent of the quantities of all other goods; that is,

(55) $\qquad\qquad \dfrac{\partial}{\partial x_t}\left[\dfrac{U_i(X)}{U_j(X)}\right] = 0, \qquad t \neq i, j.$

This necessary and sufficient condition for direct additivity can be reformulated as a condition on the preferences themselves. We begin with an example that violates direct additivity. Suppose there are three goods: bread (x_1), butter (x_2), and marmalade (x_3). An individual who lives by bread with butter and bread with marmalade, but not by bread alone, may prefer $(10, 1, 10)$ to $(5, 5, 10)$ and prefer $(5, 5, 1)$ to $(10, 1, 1)$. This pair of choices is incompatible with direct additivity. With direct additivity, if an individual prefers $(10, 1, 10)$ to $(5, 5, 10)$, then the individual will prefer $(10, 1, x_3)$ to $(5, 5, x_3)$ regardless of the common value of x_3. More generally, direct additivity implies that if an individual prefers $(x_1^a, x_2^a, \hat{x}_3)$ to $(x_1^b, x_2^b, \hat{x}_3)$ for a particular common value of x_3, say $\hat{x}_3$, then the individual prefers basket a to basket b for all common values of x_3. Direct additivity also implies analogous conditions when the common value good is x_1 or x_2 rather than x_3. Furthermore, if this "independence" condition holds for all partitions of the goods into two subsets—the common value good and the other two goods—then preferences exhibit direct additivity.[7] With three or more goods, preferences exhibit direct additivity if and only if, for all partitions of the goods into two subsets, an individual's preferences over vectors of goods that contain common values of the goods in one subset are independent of the levels of the goods in the common value subset.

With direct additivity the demand for each good is a function of the product of its own price and a single "index function" that depends on all prices and expenditure

(56a) $\qquad\qquad h^i(P, \mu) = H^i[R(P, \mu)p_i].$

---

[6]When there are only two goods, the conditions for and the implications of direct additivity are somewhat different, as Samuelson [1947, pp. 174–179] shows.

[7]Independence for one partition does not imply direct additivity but is compatible with an asymmetric preference structure such as $U(x_1, x_2, x_3) = V[V^*(x_1, x_2), x_3]$.

We restate this result in terms of normalized prices ($y_i = p_i/\mu$): with direct additivity the demand for each good is a function of the product of its own normalized price and an index function that depends on all normalized prices

(56b) $$h^i(Y) = H^i[S(Y)y_i].^8$$

The essence of this restriction is the existence of a single index function that appears in the demand function for each good. The LES is clearly of this form, since it can be written as

(57a) $$h^i(P, \mu) = b_i + \frac{a_i}{p_i}\left(\mu - \sum p_k b_k\right),$$

or, equivalently,

(57b) $$h^i(Y) = b_i + \frac{a_i}{y_i}\left(1 - \sum y_k b_k\right).$$

In the case of (57a) the index function is given by

(58a) $$R(P, \mu) = 1/[\mu - \sum p_k b_k].$$

In the case of (57b) the index function is given by

(58b) $$S(Y) = 1/[1 - \sum y_k b_k].$$

It is straightforward to show that direct additivity implies (56) and that the index function is the Lagrangian multiplier corresponding to the additive canonical form. The proof follows directly from the first order conditions for utility maximization. Using the additive canonical form, we write

(59) $$\begin{aligned} u^{i\prime}(x_i) &= \lambda y_i, \\ \sum y_k x_k &= 1. \end{aligned}$$

From these first order conditions we obtain

(60) $$x_i = H^i(\lambda y_i)$$

where $H^i(\cdot)$ is the inverse of the function $u^{i\prime}(\cdot)$ and λ is implicitly defined by the budget constraint

(61) $$\sum y_k H^k(\lambda y_k) = 1.$$

Hence, the demand functions are of the form (56b) and the index function

---

[8]Although it is sloppy notation, we use the same symbol to denote the functions $h^i(P, \mu)$ and $h^i(Y)$. It would be cleaner but more cumbersome to define a new function $h^{*i}(Y)$ by $h^{*i}(Y) = h^i(Y, 1)$, but our sloppiness should cause no confusion.

is the Lagrangian multiplier corresponding to the canonical form of the additive direct utility function.

The LES provides an instructive example. In terms of normalized prices the LES first order conditions are

(62)
$$\frac{a_i}{x_i - b_i} = \lambda y_i$$
$$\sum x_k y_k = 1.$$

Solving (62) for x_i yields

(63)
$$x_i = b_i + \frac{a_i}{\lambda y_i}$$

where λ is implicitly defined by the budget constraint

(64a)
$$1 = \sum y_k x_k = \sum y_k b_k + \frac{1}{\lambda} \sum a_k = \sum y_k b_k + \frac{1}{\lambda}.$$

With the LES we can solve explicitly for λ:

(64b)
$$\lambda = \frac{1}{1 - \sum y_k b_k},$$

which is the index function appearing in Eq. (58b).

Returning to the general additive case, we next examine the restrictions on the derivatives of the demand functions that follow from the fact that both the prices of other goods and expenditure enter the demand for a particular good only through the index function. Differentiating (56a) with respect to p_j and μ yields:

(65)
$$\frac{\partial h^i}{\partial p_j} = H^{i'} R_j p_i, \qquad i \neq j$$

and

(66)
$$\frac{\partial h^i}{\partial \mu} = H^{i'} R_\mu p_i.$$

Provided (66) is not zero, we can eliminate the common factor between (65) and (66) to obtain

(67)
$$\frac{\partial h^i}{\partial p_j} = \theta^j(P, \mu) \frac{\partial h^i}{\partial \mu}, \qquad i \neq j$$

where the factor of proportionality θ^j is given by

(68)
$$\theta^j(P, \mu) = \frac{\partial R/\partial p_j}{\partial R/\partial \mu}.$$

That is, the change in the consumption of food induced by a change in the price of clothing is proportional to the change in the consumption of food induced by a change in expenditure; the factor of proportionality depends on the good whose price has changed, but not on the good whose quantity response we are considering. We can express (67) in ratio form as

(69)
$$\frac{\partial h^i/\partial p_j}{\partial h^t/\partial p_j} = \frac{\partial h^i/\partial \mu}{\partial h^t/\partial \mu}, \qquad j \neq i, t.$$

The analysis can be extended to the partial derivatives of the compensated demand functions. The ratio of the ijth substitution term to the product of the partial derivatives with respect to expenditure of the ordinary demand functions for goods i and j is independent of i and j. More precisely

(70)
$$\frac{(\partial h^i/\partial p_j) + h^j(\partial h^i/\partial \mu)}{(\partial h^i/\partial \mu)(\partial h^j/\partial \mu)} = \Gamma(P, \mu), \qquad i \neq j.$$

To show that this ratio is independent of i and j, we substitute for $\partial h^i/\partial p_j$ from (67) and observe that the ratio is independent of i. The parallel argument using the symmetric substitution term yields a similar expression that is independent of j. Hence, the ratio must be independent of both i and j as (70) asserts.

The assumption of direct additivity or "independent utilities" has a long history in economic analysis. Samuelson [1947, p. 93] attributes the general additive form to Jevons, while crediting Gossen with the additive quadratic utility function. The class of additive direct utility functions that yields demand functions exhibiting expenditure proportionality is known as the "Bergson family," in honor of Abram Bergson (Burk [1936]). In fact, Bergson investigated a somewhat different notion of independence and did not characterize the class that now bears his name. The Bergson family consists of the CES, its Cobb–Douglas limiting case, and sometimes (depending on an author's taste in regularity conditions) its Leontief and linear limiting cases.

The class of additive direct utility functions that yields demand functions locally linear in expenditure consists of the generalized CES, its two limiting cases, the LES and the translated Leontief, and additive exponential utility functions of the form

(71a)
$$U(X) = -\sum \alpha_k e^{-\beta_k x_k}, \qquad \alpha_i > 0, \quad \beta_i > 0.$$

It is convenient to rewrite (71a) as

(71b)
$$U(X) = -\sum a_k e^{(b_k - x_k)/a_k}$$

where $a_i = 1/\beta_i$ and $b_i = (\log \alpha_i \beta_i)/\beta_i$. The corresponding demand functions are

(72)
$$h^i(P, \mu) = b_i - \frac{a_i \sum p_k b_k}{\sum p_k a_k} + \frac{a_i \mu}{\sum p_k a_k} - a_i \log p_i + \frac{a_i \sum p_k a_k \log p_k}{\sum p_k a_k}$$

and the income–consumption curves are parallel straight lines. The indirect utility function is given by the Gorman polar form, (16), where

(73) $$g(P) = \sum p_k a_k$$

and

(74) $$f(P) = (\sum p_k a_k)(\log \sum p_k a_k) - \sum p_k b_k + \sum p_k a_k \log p_k.$$

Pollak [1971b], which characterized the class of additive utility functions yielding demand functions linear in expenditure, shows that the generalized CES and the exponential cases are related through the class of utility functions

(75) $$U(X) = \sum \alpha_k (\beta_k + \delta_k x_k)^c.$$

It is easy to show that all admissible utility functions of this form yield demand systems locally linear in expenditure and that any admissible utility function of this form can be written in one of the three generalized CES forms. It is also clear that the exponential class is a limiting case of (75); more precisely,

(76) $$\lim_{c \to -\infty} - \sum \alpha_k \left(1 + \frac{-\beta_k}{c} x_k \right)^c = - \sum \alpha_k e^{-\beta_k x_k}.$$

2.2. Indirect Additivity

We say that a preference ordering is indirectly additive if it can be represented by an indirect utility function of the form

(77) $$\phi(Y) = T[\sum \phi^k(y_k)]$$

where $T(\cdot) > 0$ and the y's are normalized prices. Indirect additivity, like direct additivity, is an ordinal property. Applying Roy's identity, Eq. (19b) of Chapter 1 yields the ordinary demand functions:

(78a) $$h^i(Y) = \frac{\phi^{i'}(y_i)}{\sum y_k \phi^{k'}(y_k)} = \frac{\phi^{i'}(y_i)}{S(Y)},$$

or, in terms of unnormalized prices,

(78b) $$h^i(P, \mu) = \frac{\phi^{i'}(p_i/\mu)}{R(P, \mu)},$$

where the function R is defined in the obvious way. Indirect additivity thus implies that the demand for each good is a function of its own normalized price divided by an index function that depends on all normalized prices. Hence, the ratio of the demand functions for two goods depends on the normalized prices of the two goods but not on the index function:

(79) $$\frac{h^i(Y)}{h^j(Y)} = \frac{\phi^{i'}(y_i)}{\phi^{j'}(y_j)}.$$

Although (78) and (79) are the most transparent ways to express the implications of indirect additivity, it is straightforward to establish the consequences of indirect additivity for the partial derivatives of the demand functions.

The "indirect addilog" (Houthakker [1960]) provides a good illustration of indirect additivity. The indirect addilog utility function is given by

$$
(80) \qquad \phi(Y) = \sum \frac{\alpha_k}{\beta_k + 1} y_k^{\beta_k + 1}
$$

and the ordinary demand functions by

$$
(81) \qquad h^i(Y) = \frac{\alpha_i y_i^{\beta_i}}{\sum \alpha_k y_k^{\beta_k + 1}}.
$$

Applying to the indirect addilog utility function the analogue of the translating trick applied to the direct utility function in Eq. (18), we replace each y_i by $y_i - \gamma_i$. This yields a new indirect utility function whose origin is the point $\gamma = (\gamma_1, \ldots, \gamma_n)$:

$$
(82) \qquad \phi(Y) = \sum \frac{\alpha_k}{\beta_k + 1} (y_k - \gamma_k)^{\beta_k + 1}.
$$

The corresponding ordinary demand functions are

$$
(83) \qquad h^i(Y) = \frac{\alpha_i (y_i - \gamma_i)^{\beta_i}}{\sum \alpha_k (y_k - \gamma_k)^{\beta_k + 1}}.
$$

The CES is a special case of the indirect addilog: when all of the β's are equal to a common value, then the indirect utility function (80) becomes

$$
(84) \qquad \phi(Y) = \sum \frac{\alpha_k}{\beta + 1} y_k^{\beta + 1}
$$

and the ordinary demand functions (81) become

$$
(85) \qquad h^i(Y) = \frac{\alpha_i y_i^{\beta}}{\sum \alpha_k y_k^{\beta + 1}}.
$$

It is easy to verify that (85) is the demand system corresponding to the CES, (7).

The CES demand system is exceptional in that the direct utility function and indirect utility function that generate it have a similar mathematical form. (The underlying notion of "self-dual preferences" was introduced by Houthakker [1965] and formalized in Pollak [1972].) In general, the direct and indirect utility functions corresponding to a particular class of preference orderings have different forms. Only a narrowly circumscribed class of preference orderings exhibit both direct and indirect additivity. Hicks [1969] and Samuelson [1969] show that this class consists of the

CES, its two limiting cases, the Cobb–Douglas and the Leontief, and the preference ordering corresponding to the direct utility function

$$(86) \qquad U(X) = u^1(x_1) + \sum_{k=2}^{n} a_k \log x_k.$$

With the exception of these cases, indirect utility functions corresponding to additive direct utility functions are not additive. The reader can verify, for example, that the indirect utility functions corresponding to the LES and the generalized CES are not additive.

2.3. Generalized Additive Separability

The crucial common feature of direct additivity and indirect additivity is that the implied demand for each good can be written as a function of the normalized price of that good and a single index function that appears in every demand function. Following Pollak [1972] we say that a demand system with this feature exhibits "generalized additive separability" (GAS):

$$(87) \qquad h^i(Y) = F^i[y_i, S(Y)].$$

Conditions on the derivatives and ratios of derivatives are easily derived from (87). The class of demand systems exhibiting GAS includes not only direct additivity and indirect additivity, but also the Fourgeaud–Nataf [1959] class of demand functions

$$(88) \qquad h^i(P, \mu) = F^i\left[\frac{p_i}{R(P)}, \frac{\mu}{R(P)}\right]$$

where $R(P)$ is a function homogeneous of degree 1. The form Houthakker [1965] calls the "self-dual addilog"

$$(89) \qquad h^i(Y) = \alpha_i y_i^{\beta_i} S(Y)^{(1-\delta) + \delta\beta_i},$$

where $S(Y)$ is defined implicitly by

$$(90) \qquad \sum \alpha_k y_k^{\beta_k + 1} S(Y)^{(1-\delta) + \delta\beta_k} = 1,$$

also exhibits GAS. Using a result of Gorman, Pollak [1972] characterizes the direct and indirect utility functions corresponding to GAS.

2.4. Weak Separability and Strong Separability

A direct utility function, $U(x_1, \ldots, x_n)$ is said to exhibit "weak separability" if there exists a partition of the n goods into m subsets, m functions $V^r(X_r)$, and a function V such that

$$(91) \qquad U(X) = V[V^1(X_1), \ldots, V^m(X_m)]$$

where $m \geq 2$ and X_r is the vector of goods in the rth subset. It is convenient to use double subscripts to denote goods: the first subscript indicates the subset to which a good belongs, and the second indicates the particular good within the subset. Thus, x_{ri} denotes the ith good in the rth subset. We denote the number of goods in the rth subset by n_r. Thus, $X_r = (x_{r1}, \ldots, x_{rn_r})$ and $n_1 + n_2 + \cdots + n_m = n$.

A utility function is weakly separable if and only if the goods can be partitioned into subsets in such a way that every marginal rate of substitution involving two goods from the same subset depends only on the goods in that subset. If the utility function is weakly separable, it is easy to verify that this condition is satisfied. Leontief [1947a, 1947b] shows that this condition is also sufficient.

Weak separability has strong intuitive appeal. If we partition the goods into categories such as "food," clothing," and "recreation," it is tempting to assume that the utility function is weakly separable with respect to the subsets of this partition. (Strictly speaking, it is the set of indexes by which the commodities are identified—the integers from 1 to n—that are partitioned.) Similarly, if $U(X_1, \ldots, X_T)$ is an intertemporal utility function, where X_t denotes the vector of goods consumed in period t, it is again tempting to assume that the utility function exhibits weak separability, reflecting a type of intertemporal independence. The crucial assumption in both of these cases is that the individual's preference between two vectors of goods that differ only in the components of one subset, say "food," is independent of the identical nonfood components of the two vectors.

To establish the implications of weak separability for the ordinary demand functions, we make use of "conditional demand functions" introduced in Pollak [1969, 1971a]. Consider an individual whose preferences can be represented by a utility function $U(x_1, \ldots, x_n)$ that is not necessarily separable. If the individual's consumption of one good has been determined before he enters the market, we say that the good has been "preallocated." We assume that the individual is not allowed to sell any of his allotment of a preallocated good, and that he cannot buy more of it. For definiteness, suppose that the nth good is preallocated, while the remaining $n-1$ goods are available on the market at prices over which the individual has no control, and that his expenditure on the available goods, μ_θ, is also predetermined. The individual is supposed to choose quantities of the first $n-1$ goods so as to maximize $U(x_1, \ldots, x_n)$ subject to the "budget constraint"

$$(92) \qquad \sum_{k=1}^{n-1} p_k x_k = \mu_\theta$$

and the additional constraint

$$(93) \qquad x_n = \bar{x}_n$$

where $\bar{x}_n$ denotes his allotment of the nth good. Hence, his demand for the goods available on the market depends on the prices of these goods, total expenditure on them, and his allotment of the nth good; that is,

$$(94) \qquad x_i = g^{i \cdot n}(p_1, \ldots, p_{n-1}, \mu_\theta, x_n), \qquad i \neq n.$$

We call the function $g^{i \cdot n}(\cdot)$ the "conditional demand function" for the ith good. The second superscript, n, indicates that the nth good is pre-allocated, a terminology suggested by the analogy with conditional probability.

Conditional demand functions can also be defined when more than one good is preallocated. In general, a conditional demand function expresses the demand for a good available on the market as a function of three kinds of variables: the prices of all goods available on the market, total expenditure on these goods, and the quantities of the preallocated goods.

Formally, we partition the set of all commodities into two subsets, θ and $\bar{\theta}$. We assume that the goods in θ are available on the market, while those in $\bar{\theta}$ are preallocated; thus, if $j \in \theta$, then x_j is available on the market, while if $j \in \bar{\theta}$, then x_j is preallocated. We denote total expenditure on the goods available on the market by μ_θ. The individual is supposed to maximize $U(x_1, \ldots, x_n)$ subject to the "budget constraint"

$$(95) \qquad \sum_{k \in \theta} p_k x_k = \mu_\theta$$

and the additional constraints

$$(96) \qquad x_j = \bar{x}_j, \qquad j \in \bar{\theta}.$$

The demand for a good available on the market depends on the prices of the goods available on the market, total expenditure on them, and the quantities of the preallocated goods. Thus

$$(97) \qquad x_i = g^{i \cdot \bar{\theta}}(P_\theta, \mu_\theta, X_{\bar{\theta}}), \qquad i \in \theta,$$

where P_θ denotes the vector of prices of the goods available on the market and $X_{\bar{\theta}}$ denotes the vector of preallocated goods. The function $g^{i \cdot \bar{\theta}}$ is the conditional demand function for the ith good; the second superscript, $\bar{\theta}$, indicates that the goods in $\bar{\theta}$ are preallocated.

Now consider the relation between conditional demand functions and ordinary demand functions. Suppose that only the nth good is preallocated and that the individual's allotment of the nth good is precisely equal to the amount he would have purchased when facing prices P with expenditure μ. That is, $x_n = h^n(P, \mu)$. Suppose further that the amount he has to spend on the first $n - 1$ goods is precisely equal to the amount he would spend on these goods when facing prices P with expenditure μ: $\mu - p_n h^n(P, \mu)$. In this situation, the individual will purchase the same quantities of each of the goods available on the market as he would purchase if he faced prices P with expenditure μ and the nth good were not preallocated; that is

(98) $\qquad h^i(P, \mu) = g^{i \cdot n}[p_1, \ldots, p_{n-1}, M^\theta(P, \mu), h^n(P, \mu)], \qquad i \neq n,$

where

(99) $\qquad M^\theta(P, \mu) = \mu - p_n h^n(P, \mu) = \sum_{k=1}^{n-1} p_k h^k(P, \mu).$

This identity follows directly from the definitions of ordinary demand functions and conditional demand functions as solutions to related constrained maximization problems.

A similar result holds when more than one good is preallocated. Let $h^{\bar\theta}(P, \mu)$ denote the vector of ordinary demand functions for the preallocated goods. For example, if the last two goods are preallocated, we have $h^{\bar\theta}(P, \mu) = [h^{n-1}(P, \mu), h^n(P, \mu)]$. If all goods were available on the market, total expenditure on the goods in $\bar\theta$ would be $\sum_{k \in \bar\theta} p_k h^k(P, \mu)$ and total expenditure on the goods in θ would be

(100) $\qquad M^\theta(P, \mu) = \mu - \sum_{k \in \theta} p_k h^k(P, \mu) = \sum_{k \in \theta} p_k h^k(P, \mu).$

By the argument used to establish (98) it follows that

(101) $\qquad h^i(P, \mu) = g^{i \cdot \bar\theta}[P_\theta, M^\theta(P, \mu), h^{\bar\theta}(P, \mu)], \qquad i \in \theta,$

where M^θ is defined by (100).

Conditional demand functions, introduced in Pollak [1969], can be used to analyze consumer behavior under rationing, to provide simple proofs of the Hicksian aggregation theorem and Samuelson's LeChatelier principle, and to decompose the cross-price derivatives of the ordinary demand functions. In this book, however, we use them only to examine the implications of separability.

We now examine the conditional demand functions corresponding to a weakly separable utility function. We assume that the goods in one subset are available on the market, while all other goods are preallocated. For definiteness, suppose that the goods available on the market are in subset r, and that subset r is "food." Formally, $tj \in \theta$ if $t = r$ and $tj \in \bar\theta$ if $t \neq r$. In accordance with our previous notation, μ_θ denotes total expenditure on food and P_θ the vector of food prices. The conditional demand functions are determined by maximizing the utility function (91) subject to the budget constraint

(102) $\qquad \sum_{k=1}^{n_r} P_{rk} x_{rk} = \sum_{k \in \theta} p_k x_k = \mu_\theta$

and the additional constraints

(103) $\qquad x_{sj} = \bar{x}_{sj}, \qquad sj \in \bar\theta.$

If we absorb the constraints (103) into the utility function we obtain

$$(104) \quad V[V^1(\bar{X}_1), \ldots, V^{r-1}(\bar{X}_{r-1}), V^r(X_r), V^{r+1}(\bar{X}_{r+1}), \ldots, V^m(\bar{X}_m)].$$

Clearly, the utility maximizing values of $(x_{r1}, \ldots, x_{rn_r})$ are independent of the preallocated goods: regardless of the levels of the preallocated goods, the individual has only to maximize $V^r(X_r)$ subject to (102). Hence, the conditional demand functions for the goods in θ are of the form

$$(105) \qquad g^{ri\cdot\bar{\theta}}(P_\theta, \mu_\theta, X_{\bar{\theta}}) = g^{ri\cdot\bar{\theta}}(P_\theta, \mu_\theta).$$

It follows from the general relation between ordinary demand functions and conditional demand functions (101) that with weak separability

$$(106) \qquad h^{ri}(P, \mu) = g^{ri\cdot\bar{\theta}}[P_\theta, M^\theta(P, \mu)]$$

where M^θ is defined by (100).

This result expresses the key implication of weak separability. It implies that the demand for a good in a particular subset (e.g., Swiss cheese in the food subset) can be expressed as a function of the prices of the goods in that subset and total expenditure on those goods (food prices and total expenditure on food). Although it would be more convenient if the demand for Swiss cheese depended only on food prices and total expenditure (μ), this is not what separability implies. Instead, separability implies that expenditure and the prices of goods outside the food subset enter the demand functions for food only through their effect on total expenditure on food. Thus, if we work with expenditure on food rather than total expenditure on all goods, we can ignore the prices of goods outside the food category.

Our characterization of the ordinary demand functions implied by weak separability leads directly to several conclusions about the effects of finite price and expenditure changes. Consider the change in the consumption of Swiss cheese when there is a change in the price of some nonfood item, say, shoes; we do not require that the change in the price of shoes be small. The change in the price of shoes will cause the individual to change his total expenditure on food (μ_θ) and the change in total expenditure on food will cause a change in Swiss cheese consumption. Now consider a change in the price of another nonfood item, say, tennis balls. Suppose that the effect on μ_θ of this price change is the same as the effect on μ_θ of the change in the price of shoes. Then the change in the price of tennis balls and the change in the price of shoes will have the same effects on the consumption of Swiss cheese. Similarly, if a change in expenditure has the same effect on μ_θ as the change in the price of shoes, then it will also have the same effect on Swiss cheese consumption. A similar result holds for simultaneous changes in expenditure and in the prices of several nonfood items.

The literature on separability has been more concerned with the partial derivatives of the demand functions than with the demand functions themselves. The implications of weak separability for the partial derivatives of the demand functions follow immediately from the results we have already established. Differentiating (106) with respect to p_{sj} and μ we obtain

$$(107) \qquad \frac{\partial h^{ri}}{\partial p_{sj}} = \frac{\partial g^{ri \cdot \bar{\theta}}}{\partial \mu_{\theta}} \frac{\partial M^{\theta}}{\partial p_{sj}}, \qquad s \neq r,$$

$$(108) \qquad \frac{\partial h^{ri}}{\partial \mu} = \frac{\partial g^{ri \cdot \bar{\theta}}}{\partial \mu_{\theta}} \frac{\partial M^{\theta}}{\partial \mu}.$$

That is, the change in the consumption of Swiss cheese caused by a change in the price of shoes (expenditure) is proportional to the change in total expenditure on food caused by the change in the price of shoes (expenditure). Provided $\partial h^{ri}/\partial p_{t\ell} \neq 0$, we may express (107) in ratio form as

$$(109) \qquad \frac{\partial h^{ri}/\partial p_{sj}}{\partial h^{ri}/\partial p_{t\ell}} = \frac{\partial M^{\theta}/\partial p_{sj}}{\partial M^{\theta}/\partial p_{t\ell}}, \qquad s, t \neq r.$$

Provided $\partial h^{ri}/\partial \mu \neq 0$, we can eliminate $\partial g^{ri \cdot \bar{\theta}}/\partial \mu$ between (107) and (108) and obtain

$$(110) \qquad \frac{\partial h^{ri}}{\partial p_{sj}} = v_r^{sj} \frac{\partial h^{ri}}{\partial \mu}, \qquad s \neq r,$$

where v_r^{sj} is defined by

$$(111) \qquad v_r^{sj} = \frac{\partial M^{\theta}/\partial p_{sj}}{\partial M^{\theta}/\partial \mu}.$$

That is, the change in the consumption of Swiss cheese induced by a change in the price of shoes is proportional to the change in the consumption of Swiss cheese induced by a change in expenditure. The factor of proportionality is not a constant, but a function of all prices and expenditure. It is the same for all food items (Swiss cheese, roast beef), but it does depend on the good whose price has changed. We can express (110) in ratio form as

$$(112) \qquad \frac{\partial h^{ri}/\partial p_{sj}}{\partial h^{r\ell}/\partial p_{sj}} = \frac{\partial h^{ri}/\partial \mu}{\partial h^{r\ell}/\partial \mu}, \qquad s \neq r.$$

We now turn from weak separability to "strong separability." A direct utility function is said to exhibit strong separability if there exists a partition of the commodities into m subsets, m functions $V^r(X_r)$, and a function T, $T'(\cdot) > 0$, such that

$$(113) \qquad U(X) = T\left[\sum_{r=1}^{m} V^r(X_r) \right].$$

If $m \geqslant 3$, a utility function is strongly separable if and only if the goods can be partitioned into subsets in such a way that every marginal rate of substitution involving goods from different subsets depends only on the goods in those two subsets. Because a utility function that is strongly separable with m subsets is also weakly separable with m subsets, our weak separability results apply directly to strong separability. Hence, we discuss only the additional implications of strong separability.

A weakly separable utility function is also called a "utility tree" and the subsets "branches." Strong separability is sometimes called "block additivity" and the subsets "blocks." This terminology is useful in discussing the additional implications of strong separability. If a utility function is a tree with m branches, in general we cannot combine two branches into a single branch and treat the utility function as a tree with m − 1 branches. That is, with weak separability the marginal rate of substitution between Swiss cheese and tennis balls can depend on the quantity of shoes. With block additivity, however, it is always permissible to treat the goods in two (or more) blocks as a single blocks. For example, to combine the goods in blocks s and t into a single block, we first renumber the blocks so that s = 1 and t = 2, and then note that (113) can be rewritten as

$$(114) \qquad U(X) = T\left[V^1(X_1) + V^2(X_2) + \sum_{r=3}^{m} V^r(X_r) \right].$$

Hence, if the utility function is strongly separable, we may write the ordinary demand functions as

$$(115) \qquad h^{ri}(P, \mu) = g^{ri \cdot \bar{\theta}}[P_r, P_t, M^\theta(P, \mu)],$$

where θ denotes the set of all goods in blocks r and t, and $\bar{\theta}$ the goods in the remaining blocks. The implications of strong separability that go beyond those of weak separability can easily be established by combining blocks into superblocks and recognizing that the utility function is weakly separable in those superblocks. (When there are only two blocks this technique does not yield any implications of strong separability that go beyond those of weak separability. Recall that direct additivity has different implications for cases with two goods and cases with three or more goods.)

To illustrate this superblock technique, consider the case in which $\bar{\theta}$ denotes the goods in one block and θ the goods in the remaining m − 1 blocks. For definiteness, suppose that the goods in block s are in $\bar{\theta}$ and the remaining goods are in θ. Then

$$(116) \qquad h^{ri}(P, \mu) = g^{ri \cdot \bar{\theta}}[P_\theta, M^\theta(P, \mu)], \qquad r \neq s.$$

If we take block s to be clothing, then (116) implies that the demand for Swiss cheese can be expressed as a function of the prices of all nonclothing goods and total expenditure on all goods other than clothing. It is not clear that it is more useful to write the demand for Swiss cheese this way than as a function of all food prices and total expenditure on food; if the utility function is strongly separable, however, then we have the option of writing the demand functions in either form and the choice between them must depend on the problem at hand. If we are primarily interested in the effect of various price and expenditure changes on Swiss cheese consumption, then it is probably more convenient to write the demand for Swiss cheese as a function of all food prices and total expenditure on food. If, on the other hand, we are primarily interested in the effect of a change in the price of shoes on the consumption of a variety of nonclothing goods (e.g., Swiss cheese, tennis balls, etc.), then it is more convenient to write the demands for these goods as functions of all nonclothing prices and total expenditure on goods outside the clothing category. The implications of (116) for finite price and expenditure changes should be clear from our earlier discussion of weak separability. The derivation of the implications for the partial derivatives follows the same general route and yields

$$(117) \qquad \frac{\partial h^{ri}}{\partial p_{sj}} = v^{sj} \frac{\partial h^{ri}}{\partial \mu}, \qquad s \neq r$$

where v^{sj} is defined by

$$(118) \qquad v^{sj} = \frac{\partial M^{\theta}/\partial p_{sj}}{\partial M^{\theta}/\partial \mu}.$$

In contrast to the tree case, (110), with block additivity the factor of proportionality is independent of r.

Although we examined the implications of direct additivity in Section 2.1 we return to it here because of its relation to weak and strong separability. To avoid the exceptional two-good case, we assume $n \geqslant 3$. If an individual's utility function is strongly separable with blocks corresponding to "food," "clothing," etc., and if we form the Hicksian composite commodities corresponding to these blocks, then the utility function defined in terms of these composite commodities is additive. In describing the implications of additivity, we shall refer to the commodities as "food," "clothing," etc.

If a utility function is additive, then it is also strongly separable; for regardless of how the goods are partitioned into subsets, they will satisfy (113). This means that the results for weak and strong separability can be used to deduce the implications of additivity. Let θ and $\bar{\theta}$ be any partition of the goods into two subsets. Then the demand for the goods in θ can be written as a function of the prices of the goods in θ and total expenditure on these goods. This result holds for all possible partitions of the commodities into subsets and, thus, when the utility function is additive, we may use any partition of the goods that is convenient. For example, we can write the demand for food as a function of the price of food, the price of recreation, and total expenditure on food and recreation:

$$(119) \qquad h^r(P,\mu) = g^{r\cdot\bar\theta}[p_r, p_t, M^\theta(P,\mu)]$$

where $\theta = \{r, t\}$ and all other goods are in $\bar\theta$. Or, if we prefer, we can write the demand for food as a function of the prices of all goods except clothing and total expenditure on all goods other than clothing:

$$(120) \qquad h^r(P,\mu) = g^{r\cdot\bar\theta}[P_\theta, M^\theta(P,\mu)],$$

where $\theta = \{1,\ldots,s-1, s+1,\ldots,n\}$ and $\bar\theta = \{s\}$. The implications of these results for finite price and expenditure changes are obvious and we shall not discuss them. The implications for the partial derivatives can be derived by differentiating (120) with respect to p_s and μ; ratios of these expressions yield results analogous to those obtained under weak and strong separability and provide an alternative derivation of (69).

In our discussion of additivity in Section 2.1 we derived an expression involving the partial derivatives of the compensated demand functions, (70). Under weak separability, we can obtain an analogous result: if a utility function is a tree, then there exist $(n^2 - n)/2$ functions $\{\Gamma^{rs}(P,\mu)\}, \Gamma^{rs}(P,\mu) = \Gamma^{sr}(P,\mu)$, such that

$$(121) \qquad \frac{(\partial h^{ri}/\partial p_{sj}) + h^{sj}(\partial h^{ri}/\partial\mu)}{(\partial h^{ri}/\partial\mu)(\partial h^{sj}/\partial\mu)} = \Gamma^{rs}(P,\mu) = \Gamma^{sr}(P,\mu), \qquad s \neq r.$$

Goldman and Uzawa [1964] show that this condition is necessary and sufficient for weak separability; we shall prove only necessity. To show that this ratio is independent of i and j, we substitute from (110) into the left-hand side of (121) and obtain an expression that depends on sj and r, but is independent of i. The parallel argument using the symmetric substitution term yields a similar expression that is independent of j. Hence, the ratio must be independent of both i and j, although it does depend on both r and s. We denote the common value of these ratios by $\Gamma^{rs}(P,\mu) = \Gamma^{sr}(P,\mu)$.

If the utility function is strongly separable, the argument used in the weakly separable case to establish (121) can be used to show

$$(122) \qquad \frac{(\partial h^{ri}/\partial p_{sj}) + h^{sj}(\partial h^{ri}/\partial\mu)}{(\partial h^{ri}/\partial\mu)(\partial h^{sj}/\partial\mu)} = \Gamma(P,\mu), \qquad s \neq r.$$

That is, instead of $(n^2 - n)/2$ functions $\{\Gamma^{rs}(P,\mu)\}$, there is only a single function, $\Gamma(P,\mu)$.

The "S-branch utility function" of Brown and Heien [1972] provides an instructive example of a specification suitable for empirical work. They postulate a strongly separable utility function in which the block utility functions are of the generalized CES form and the aggregator utility function is CES. For example, the generalized CES forms (21) can be written as

(123)
$$V^r(X_r) = \left[\sum_{k=1}^{n_r} a_{rk}(x_{rk} - b_{rk})^{c_r} \right]^{1/c_r},$$

and the CES aggregator utility function as

(124)
$$V[V^1(X_1), \dots, V^m(X_m)] = \left[\sum_{r=1}^{m} a_r(V^r(X_r))^c \right]^{1/c}.$$

Letting $\sigma_r = 1/(c_r - 1)$ and $\sigma = 1/(c - 1)$, the corresponding demand system is given by

(125)
$$x_{ri} = h^{ri}(P_r, \mu_r) = b_{ri} - \gamma^{ri}(P_r) \sum p_{rk} b_{rk} + \gamma^{ri}(P_r)\mu_r,$$

where

(126)
$$\gamma^{ri}(P_r) = \frac{(p_{ri}/a_{ri})^{\sigma_r}}{\sum_{k=1}^{n_r} p_{rk}(p_{rk}/a_{rk})^{\sigma_r}}$$

and

(127)
$$\mu_r = \frac{a_r^\sigma \left[\sum_{k=1}^{n_r} a_{rk}^{\sigma_r} p_{rk}^{1-\sigma_r} \right]^{(1-\sigma)/(1-\sigma_r)}}{\sum_{s=1}^{n} a_s^\sigma \left(\sum_{k}^{n_s} a_{sk}^{\sigma_s} p_{sk}^{1-\sigma_s} \right)^{(1-\sigma)/(1-\sigma_s)}} \left(\mu - \sum_{s=1}^{m} \sum_{k=1}^{n_s} p_{sk} b_{sk} \right) + \sum_{k=1}^{n_r} p_{rk} b_{rk}.$$

The demand system is, of course, linear in expenditure. The expression for μ_r is most easily derived by substituting the indirect utility functions for the generalized CES into (124).

It is tempting to consider decomposing estimation of the S-branch into two stages, at the first stage estimating the parameters of the m conditional generalized CES demand systems corresponding to each block, and at the second stage estimating the m parameters of the CES aggregator utility function. Unfortunately, the validity of this two-stage approach to estimation depends on postulating an implausible stochastic structure that prevents disturbances associated with the demand functions for goods in one block from affecting the demand for goods in other blocks.

We conclude this section with a brief note on the literature. Leontief [1947a, 1947b] investigated the underlying mathematical structure of separability. Debreu [1960] provided an important characterization of additivity. The notions of weak and strong separability were developed in Strotz [1957, 1959] and Gorman [1959]. Goldman and Uzawa [1964] developed a characterization of separability in terms of the partial derivatives of the demand functions. Gorman [1968] provides a definitive discussion of separability concepts; even the reader who is not interested in the technical details will enjoy the subsequent exchange between Vind

[1971a, 1971b] and Gorman [1971a, 1971b]. Blackorby, Primont, and Russell [1978] provide a thorough discussion and rigorous analysis of separability.

3. FLEXIBLE FUNCTIONAL FORMS

Over the last decade flexible functional forms have come to play an increasingly important role in empirical demand analysis. The translogs are the most widely used family of flexible functional forms, and we begin this section by providing an extensive discussion of translog specifications. This marks a departure from our usual practice of emphasizing broad classes of demand systems rather than particular specifications. We deviate in this case not only because of the importance of translog specifications, but also because we think that translogs, like many other topics, are "best understood by staring hard at some non-trivial examples" (Halmos [1958, p. 42]). In Section 3.2 we use the translog family as the starting point for a more general discussion of flexible functional forms. Finally, in Section 3.3 we discuss the advantages and limitations of flexible specifications in empirical demand analysis.

3.1. The Translog Family

The translog, the most widely used family of flexible functional forms, provides a good introduction to the entire class. The "homothetic translog" (HTL), the simplest member of the translog family, is the best starting point. The HTL indirect utility function is given by

$$(128) \qquad \psi(P, \mu) = \log \mu - \sum \alpha_k \log p_k - \tfrac{1}{2} \sum_j \sum_k \beta_{kj} \log p_k \log p_j,$$

$$\beta_{ij} = \beta_{ji} \text{ for all } i,j, \qquad \sum \beta_{ki} = 0 \text{ for all } i, \qquad \sum \alpha_k = 1.$$

Recall our notational convention that a summation symbol without an index of summation indicates a sum over the index k. Hereafter, we shall use double summation symbols without indexes of summation, $\sum\sum$, to indicate sums over the indexes j and k; such double sums appear often in translogs and other flexible functional forms. Such forms often require symmetry restrictions on the elements of a coefficient matrix; hereafter, when we write $\beta_{ij} = \beta_{ji}$, we do so with the understanding that it holds for all i,j. Similarly, a requirement such as $\sum \beta_{ki} = 0$ is understood to hold for all i.

When working with translog demand systems, it is convenient to rewrite Roy's identity in share form:

$$(129) \qquad \omega^i(P, \mu) = - \frac{\partial \psi(P, \mu)/\partial \log p_i}{\partial \psi(P, \mu)/\partial \log \mu}.$$

It is also convenient to make use of the following mathematical fact, which the reader can easily verify: if $\beta_{kj} = \beta_{jk}$, then

$$(130) \qquad \frac{\partial}{\partial \log p_i} \left[\frac{1}{2} \sum\sum \beta_{kj} \log p_k \log p_j \right] = \sum_j \beta_{ij} \log p_j.$$

Using (129) and (130), it is straightforward to show that the HTL share equations are given by

$$(131) \qquad \omega^i(P, \mu) = \alpha_i + \sum_j \beta_{ij} \log p_j.$$

As its name indicates, the HTL corresponds to homothetic preferences and, hence, the corresponding demand system exhibits expenditure proportionality. Although this property makes the HTL uninteresting for empirical demand analysis, we discuss it here for three reasons. First, because the HTL is the simplest translog, it provides a good introduction to the entire family. Second, the HTL is a useful building block for constructing interesting consumer demand systems: it is, as we shall see, the common element that unites the translog family. Third, the HTL plays an important role in the analysis of producer behavior. The HTL indirect utility function, (128), can be solved explicitly for the log expenditure function or, in the terminology of production theory, the log cost function

$$(132) \qquad \log \mu = \log q + \sum \alpha_k \log p_k + \frac{1}{2} \sum\sum \beta_{kj} \log p_k \log p_j,$$

where q is output and μ is total cost. This translog cost function, suitably generalized to allow nonconstant returns to scale and technical progress, is the most popular specification in the empirical analysis of cost functions and factor demand systems.

The HTL, like all members of the translog family, is a generalization of the Cobb–Douglas and reduces to it when all of the β's are 0. The α's of the HTL have a straightforward behavioral interpretation: when all prices are unity, the α's are equal to the expenditure shares. As we show below, the β's also can be given a straightforward behavioral interpretation.

The linear translog (LTL) indirect utility function is given by

$$(133) \quad \psi(P, \mu) = \log(\mu - \sum p_k b_k) - \sum \alpha_k \log p_k - \frac{1}{2} \sum\sum \beta_{kj} \log p_k \log p_j,$$

$$\beta_{ij} = \beta_{ji}, \qquad \sum \beta_{ki} = 0, \qquad \sum \alpha_k = 1.$$

The LTL is obtained by translating the origin of the HTL, a process that might loosely be described as introducing "committed quantities" into the homothetic system. The LTL share equations,

$$(134) \qquad \omega^i(P, \mu) = \frac{p_i b_i}{\mu} + \left[\alpha_i + \sum_j \beta_{ij} \log p_j \right] \left[1 - \sum \frac{p_k}{\mu} b_k \right],$$

correspond to a demand system linear in expenditure. Both the LES and the HTL are special cases of the LTL. The LTL, the first translog form in the literature, was proposed by Lau and Mitchell (1971] and estimated by Manser [1976].

The log translog (log TL) indirect utility function is given by

$$(135) \qquad \psi(P, \mu) = - \sum \alpha_k \log(p_k/\mu) - \frac{1}{2} \sum\sum \beta_{kj} \log(p_k/\mu) \log(p_j/\mu),$$

$$\beta_{ij} = \beta_{ji}, \qquad \sum\sum \beta_{kj} = 0, \qquad \sum \alpha_k = 1,$$

and the corresponding share equations by

$$(136) \qquad \omega^i(P, \mu) = \frac{\alpha_i + \sum\limits_j \beta_{ij} \log p_j - \log \mu \sum\limits_j \beta_{ij}}{1 + \sum\sum \beta_{kj} \log p_j}.$$

Because the log TL share equations are linear in the log of expenditure, the log TL belongs to the PIGLOG class. The log TL has recently played a major role in demand system estimation; see, for example, Jorgenson [1990]. In the literature both the log TL—the term is ours—and a form we call the "basic translog" (BTL) are often referred to as "the translog," a practice virtually guaranteed to cause confusion.

The AIDS demand system of Deaton and Muellbauer [1980a]—the unfortunate acronym resulted from their decision to give the system a persuasive name ("Almost Ideal Demand System") rather than a descriptive one—is, like the log TL, a member of both the translog family and the PIGLOG class. We saw in Section 1.4 that the PIGLOG demand systems are generated by indirect utility functions of the form

$$(137) \qquad \psi(P, \mu) = G(P)[\log \mu - \log g(P)]$$

where $G(\lambda P) = G(P)$ and $g(\lambda P) = \lambda g(P)$ and that the share equations are given by

$$(138) \qquad \omega^i(P, \mu) = \frac{p_i g_i}{g} - \frac{p_i G_i}{G}[\log \mu - \log g].$$

The AIDS system is the special case in which

$$(139) \qquad G(P) = \prod p_k^{-\gamma_k}, \qquad \sum \gamma_k = 0$$

and

$$(140) \qquad \log g(P) = \alpha_0 + \sum \alpha_k \log p_k + \frac{1}{2} \sum\sum \beta_{kj} \log p_k \log p_j,$$

$$\beta_{ij} = \beta_{ji}, \qquad \sum \beta_{ki} = 0, \qquad \sum \alpha_k = 1.$$

It is easy to verify that the AIDS share equations are given by

$$(141) \qquad \omega^i(P, \mu) = \alpha_i + \sum \beta_{ki} \log p_k + \gamma_i[\log \mu - \log g].$$

The AIDS demand system reduces to the HTL when all of the γ's are 0.

The "basic translog" (BTL) indirect utility function, introduced by Christensen, Jorgenson, and Lau [1975], is given by

$$(142) \qquad \psi(P,\mu) = -\sum \alpha_k \log(p_k/\mu) - \frac{1}{2}\sum\sum \beta_{kj} \log(p_k/\mu) \log(p_j/\mu),$$

$$\beta_{ij} = \beta_{ji}, \qquad \sum \alpha_k = 1.$$

It is straightforward to show that the BTL share equations are given by

$$(143a) \qquad \omega^i(P,\mu) = \frac{\alpha_i + \sum_j \beta_{ij} \log(p_j/\mu)}{1 + \sum\sum \beta_{kj} \log(p_j/\mu)},$$

or, equivalently,

$$(143b) \qquad \omega^i(P,\mu) = \frac{\alpha_i + \sum_j \beta_{ij} \log p_j - \log \mu \sum_j \beta_{ij}}{1 + \sum\sum \beta_{kj} \log p_j - B \log \mu},$$

where

$$(144) \qquad\qquad\qquad B = \sum\sum \beta_{kj}.$$

(The normalization rule $\sum \alpha_k + \sum\sum \beta_{kj} = 1$ is sometimes more convenient than $\sum \alpha_k = 1$; when $\sum\sum \beta_{kj} = 0$, the two normalization rules are, of course, equivalent.)

The log TL is a special case of the BTL obtained by imposing the constraint

$$(145) \qquad\qquad\qquad B = \sum\sum \beta_{kj} = 0,$$

so that expenditure drops out of the denominator of the share equation. The HTL is obtained from the BTL by imposing n parameter restrictions

$$(146) \qquad\qquad \sum_j \beta_{ij} = 0, \qquad i = 1, \ldots, n.$$

With these n restrictions the expenditure term and the price terms drop out of the denominator of the share equations, and the expenditure term drops out of the numerator.

The "generalized translog" (GTL) is obtained by translating the BTL, in the same way that the LTL was obtained by translating the HTL. The indirect utility function is given by

$$(147) \quad \psi(P,\mu) = -\sum \alpha_k \log\left[p_k \bigg/ \left(\mu - \sum_t p_t b_t \right) \right]$$

$$-\frac{1}{2}\sum\sum \beta_{kj} \log\left[p_k \bigg/ \left(\mu - \sum_t p_t b_t \right) \right] \log\left[p_j \bigg/ \left(\mu - \sum_t p_t b_t \right) \right],$$

$$\beta_{ij} = \beta_{ji}, \qquad \sum \alpha_k = 1,$$

and the corresponding share system by

(148)

$$\omega^i(P,\mu) = \frac{b_i p_i}{\mu} + \left[1 - \left(\sum p_k b_k\right)\Big/\mu\right] \frac{\alpha_i + \sum_j \beta_{ij} \log\left[p_j\Big/\left(\mu - \sum p_k b_k\right)\right]}{\sum \alpha_k + \sum\sum \beta_{jk} \log\left[p_j\Big/\left(\mu - \sum p_k b_k\right)\right]}.$$

The GTL includes as special cases both the BTL (when all of the b's are 0) and the LTL when ($\sum \beta_{ki} = 0$ for all i).

The six forms we have discussed thus far are "indirect translog" forms: they correspond to preference orderings represented by indirect utility functions and they contain a double sum over the logs of prices. In addition to the indirect translogs, the literature on consumer and producer behavior contains a number of distinct translog forms whose common feature is that they somewhere contain a double sum over the logs of prices or quantities. A representative of the other branch of the translog family competing for the consumer behavior niche is the "direct basic translog," a preference ordering whose direct utility function is given by

(149) $$U(X) = \sum a_k \log x_k + \frac{1}{2}\sum\sum b_{kj} \log x_k \log x_j,$$

$$b_{ij} = b_{ji}, \qquad \sum a_k = 1.$$

Except in degenerate special cases, the direct translog and the indirect translog correspond to different preference specifications—hardly a surprise since, except in degenerate special cases, direct additivity and indirect additivity also correspond to distinct preference specifications. Because Roy's identity enables us to derive closed-form expressions for the share equations from indirect utility functions, indirect translogs are much more widely used than direct translogs.

Two translog forms are widely used in the study of producer behavior: the translog cost function, which we have already mentioned, and the translog profit function. Diewert [1982] provides extensive references to the literature. Although the translog profit function and the translog cost function have similar mathematical structures and are both flexible functional forms, they correspond to distinct specifications of the underlying technology.

It has been suggested that a major drawback of the translog, compared, for example, to the LES, is that the translog parameters lack a behavioral interpretation. To meet this objection, we now offer an interpretation of the translog β's. Before doing so, however, it is useful to present some preliminary results that enable us to express the Slutsky symmetry conditions in terms of elasticities and to express elasticities in terms of the derivatives of the share equations. The results in the following paragraph are general ones and do not assume that the underlying demand system is a translog.

As in Chapter 1, we denote the elasticity of the demand for the ith good with respect to the price of the jth good by E^i_j and its elasticity with respect to expenditure by E^i_μ

$$(150) \qquad E^i_j = \frac{p_j}{h^i} \frac{\partial h^i}{\partial p_j}$$

$$(151) \qquad E^i_\mu = \frac{\mu}{h^i} \frac{\partial h^i}{\partial \mu}.$$

It is easy to verify that the Slutsky symmetry condition

$$(152a) \qquad \frac{\partial h^i}{\partial p_j} + h^j \frac{\partial h^i}{\partial \mu} = \frac{\partial h^j}{\partial p_i} + h^i \frac{\partial h^j}{\partial \mu}$$

can be rewritten in terms of elasticities as

$$(152b) \qquad w_i E^i_j + w_i w_j E^i_\mu = w_j E^j_i + w_i w_j E^j_\mu$$

or, equivalently,

$$(152c) \qquad w_i E^i_j - w_i w_j E^j_\mu = w_j E^j_i - w_i w_j E^i_\mu.$$

The derivatives of the share equations with respect to the log price and log expenditure are closely related to elasticities. It is straightforward to verify that, for any demand system,

$$(153) \qquad \frac{\partial \omega^i(P, \mu)}{\partial \log p_i} = w_i(1 + E^i_i),$$

$$(154) \qquad \frac{\partial \omega^i(P, \mu)}{\partial \log p_j} = w_i E^i_j, \qquad i \ne j,$$

$$(155) \qquad \frac{\partial \omega^i(P, \mu)}{\partial \log \mu} = w_i E^i_\mu - w_i.$$

With these results in hand, we turn to the analysis of the translog.

We begin with the HTL because it yields strong simple results. These results are interesting because of the HTL's importance in empirical production analysis and because they generalize to other members of the translog family. Calculating the derivatives of the HTL share equations (131) with respect to $\log p_i$ and $\log p_j$, (153) and (154) imply

$$(156) \qquad w_i(1 + E^i_i) = \beta_{ii}$$

$$(157) \qquad w_i E^i_j = \beta_{ij}, \qquad i \ne j.$$

Because of the symmetry of the β's, (157) implies

$$(158) \qquad w_i E^i_j = w_j E^j_i.$$

Differentiating the HTL share equations with respect to log μ, (155) implies

(159) $$E_\mu^i = E_\mu^j = 1,$$

a result that could have been derived directly from homotheticity. These results enable us to offer the following interpretation of the β's in the HTL: the off-diagonal β's are the product of the shares and the cross-price elasticities; the diagonal β's are the product of the shares and 1 plus the own-price elasticities. These relationships between the β's and the elasticities hold at all price–expenditure situations, not just when all prices and expenditure are unity.

The log TL and BTL yield more complex results which we derive in Appendix B. For the log TL

(160) $$\frac{\beta_{ii}}{D(P,\mu)} = w_i[(1 + E_i^i) - w_i(E_\mu^i - 1)]$$

(161) $$\frac{\beta_{ij}}{D(P,\mu)} = w_i E_j^i + w_i w_j - w_i w_j E_\mu^j, \qquad i \neq j,$$

where $D(P,\mu)$ is the expression that appears in the denominator of the log TL and BTL share equations:

(162) $$D(P,\mu) = 1 + \sum\sum \beta_{kj} \log p_j - B \log \mu.$$

For the log TL, $B = 0$ so the final term is 0. When all prices and expenditure are unity, $D(P,\mu) = 1$ and (160) and (161) provide a behavioral interpretation of the β's in the log TL.

The BTL is more complicated. In Appendix B we define the function $\varepsilon(P,\mu)$ by $\varepsilon(P,\mu) = 2B/D(P,\mu)$ and show that

(163) $$\varepsilon(P,\mu) = \frac{\partial(w_i E_\mu^i - w_i)/\partial \log \mu}{w_i E_\mu^i - w_i} = \frac{\partial^2 \omega^i(P,\mu)/\partial(\log \mu)^2}{\partial \omega^i(P,\mu)/\partial \log \mu} = \frac{2B}{D(P,\mu)}.$$

That is, the last two expressions in (163) are independent of i. We then show that

(164) $$\frac{\beta_{ii}}{D(P,\mu)} = w_i[(1 + E_i^i) - w_i(E_\mu^i - 1)] + \tfrac{1}{2} w_i^2 \varepsilon(P,\mu)$$

(165) $$\frac{\beta_{ij}}{D(P,\mu)} = w_i E_j^i + w_i w_j - w_i w_j E_\mu^j + \tfrac{1}{2} w_i w_j \varepsilon(P,\mu), \qquad i \neq j.$$

When all prices and expenditure are unity, these expressions provide a behavioral interpretation of the β's in the BTL, since the ε on the right hand side can be evaluated directly from the demand system using either of the last two expressions in (163).

3.2. Flexible Functional Forms

A demand system is said to be a "flexible functional form" if it is capable of providing a second order approximation to the behavior of any theoretically plausible demand system at a point in the price–expenditure space. More precisely, a flexible functional form can mimic not only the quantities demanded, the income derivatives, and the own-price derivatives, but also the cross-price derivatives at a particular point; equivalently, a flexible functional form can replicate not only the shares, the income elasticities, and the own-price elasticities, but also the cross-price elasticities at a specified price–expenditure situation. To understand the meaning of this definition, it is useful to begin by counting the number of shares and elasticities that a flexible function form must reproduce.

Consider an otherwise arbitrary theoretically plausible demand system and select a "point of approximation" in the price–expenditure space. At the point of approximation the demand system has n shares, n income elasticities, n own-price elasticities, and $n(n - 1)$ cross-price elasticities. In a theoretically plausible demand system, however, not all of these $n^2 + 2n$ values are independent. In particular,

- the shares must sum to unity, so only $n - 1$ of them are independent;
- given the shares, only $n - 1$ of the income elasticities are independent, since they must satisfy

$$(166) \qquad \sum w_k E_\mu^k = 1,$$

  a relation easily derived by differentiating the budget constraint with respect to μ;
- given the shares and the income elasticities, only $n(n - 1)/2$ of the cross-price elasticities are independent: given the "below diagonal" elasticities (i.e., E_j^i for $j < i$), the shares, and the income elasticities, we can infer the above diagonal elasticities using the Slutsky symmetry condition [in elasticity form, (152c)]:
- given the shares, the income elasticities, and the cross-price elasticities, all of the own-price elasticities are uniquely determined; we can infer them all, making use of the fact that the elasticities must satisfy

$$(167) \qquad w_i + \sum w_k E_i^k = 0,$$

  a relation easily derived from the budget constraint by differentiating with respect to p_i.

Adding up these numbers, we find that there are at most $n(n - 1)/2 + 2n - 2$ independent shares and elasticities in a theoretically plausible demand system. We demonstrate below that a theoretically plausible demand system can in fact have this number of independent shares and elasticities.

A functional form is flexible with respect to the class of homothetic theoretically plausible demand systems if it is capable of providing a second order approximation to the behavior of any homothetic theoretically plausible demand system at a point in the price–expenditure space. Thus, a functional form that is flexible in this restricted sense need not be able to mimic arbitrary income elasticities. It must be possible for all of the income elasticities in such a system to equal unity, but it need not be possible for them to assume any other value. A slight modification of the argument used to show that a theoretically plausible demand system contains at most $n(n-1)/2 + 2n - 2$ independent shares and elasticities shows that a *homothetic* theoretically plausible demand system contains at most $n(n-1)/2 + n - 1$ independent shares and elasticities.

The HTL provides a transparent example of a functional form that is flexible with respect to the class of homothetic theoretically plausible demand systems and thus provides a useful starting point for discussing flexible functional forms. We begin by noting that the HTL contains $n(n-1)/2 + n - 1$ independent parameters: $n - 1$ α's (since they must sum to 1), n diagonal β's, and $n(n-1)/2$ off-diagonal β's, all subject to the n constraints of Eq. (146).

To show that the HTL can mimic the share and elasticity values of any homothetic theoretically plausible demand system, we begin with any such demand system and select a point of approximation. Without loss of generality, we can redefine the units in which the goods are measured so that in the redefined units all prices are unity at the point of approximation. (When we consider the nonhomothetic log TL and BTL, we shall require all normalized prices to be unity at the point of approximation; this amounts to requiring expenditure as well as all prices to be unity.) We denote the shares at the point of approximation by $\{\bar{w}_i\}$, the income elasticities that are, of course, unity by $\{\bar{E}^i_\mu\}$, and the price elasticities by $\{\bar{E}^i_j\}$. We now choose the HTL parameters as follows: let the α's equal the shares at the point of approximation; let the β's take the values implied by Eqs. (156) and (157):

(168) $$\beta_{ii} = \bar{w}_i(1 + \bar{E}^i_i)$$

(169) $$\beta_{ij} = \bar{w}_i\bar{E}^i_j, \qquad i \neq j.$$

We now show that the β's chosen in this way satisfy the symmetry condition, $\beta_{ij} = \beta_{ji}$, and Eq. (146). Symmetry of the β's requires (158), but this condition is automatically satisfied for any system exhibiting expenditure proportionality; it follows immediately from the Slutsky conditions (152b), in conjunction with (159). It is also easy to verify that

β's chosen in this way satisfy Eq. (146), since

(170)
$$\sum_j \beta_{ij} = \sum_j \beta_{ji} = \sum_j \bar{w}_j \bar{E}_i^j + \bar{w}_i = 0$$

where the final step depends on Eq. (167).

With the HTL parameters chosen in this way, it is straightforward to verify that the HTL demand system will replicate the shares and the price elasticities (as well as the unitary income elasticities) of the original homothetic theoretically plausible demand system at the point of approximation.

The log TL is a flexible functional form, capable of mimicking arbitrary income elasticities $\{\bar{E}_\mu^i\}$ as well as arbitrary shares and the price elasticities of a theoretically plausible demand system at the point of approximation. Thus, is is not surprising that the log TL has $n(n-1)/2 + 2n - 2$ independent parameters. (To see this, begin with the HTL count and note that the log TL contains $n-1$ additional parameters: the β's need not satisfy the n constraints (146), but only the single adding-up condition (145).) To show that the log TL is flexible, we choose the parameters as follows: let the α's equal the shares at the point of approximation; let the β's assume the values specified in Eqs. (160) and (161), (where $D = 1$):

(171)
$$\beta_{ii} = \bar{w}_i + \bar{w}_i \bar{E}_i^i + \bar{w}_i^2 - \bar{w}_i^2 \bar{E}_\mu^i$$

(172)
$$\beta_{ij} = \bar{w}_i \bar{E}_j^i + \bar{w}_i \bar{w}_j - \bar{w}_i \bar{w}_j \bar{E}_\mu^j, \qquad i \neq j.$$

We must show that the β's specified in this way satisfy the symmetry requirements, $\beta_{ij} = \beta_{ji}$, and that they satisfy the adding-up condition (145). Symmetry of the β's follows from the Slutsky symmetry conditions, most conveniently expressed in (152c). To show that the adding-up condition holds, we sum (171) and (172) over all j and find

(173)
$$\sum_j \beta_{ij} = 2\bar{w}_i + \bar{w}_i \sum_j \bar{E}_j^i - \bar{w}_i \sum_j \bar{w}_j \bar{E}_\mu^j = \bar{w}_i - \bar{w}_i \bar{E}_\mu^i$$

where the final step uses (166) and the fact that the demand functions are homogeneous of degree 0 in prices and expenditure which implies

(174)
$$\sum_j E_j^i + E_\mu^i = 0.$$

Summing (173) over i yields

(175)
$$\sum_i \sum_j \beta_{ij} = 1 - \sum_i \bar{w}_i \bar{E}_\mu^i = 0$$

where the final step again makes use of (166). It is straightforward to verify that the log TL with the parameters chosen in this way mimics the behavior of the original demand system at the point of approximation.

The log TL is a parsimonious flexible functional form in the sense that it contains the minimum number of parameters required to achieve flexibility. Lau [1986, p. 1546] defines and discusses parsimony as well as other properties of flexible functional forms. The parsimony of the log TL follows immediately from a parameter count: a flexible functional form must contain at least $n(n-1)/2 + 2n - 2$ independent parameters, so any flexible form containing precisely this number of independent parameters is parsimonious.[9]

The BTL has $n(n-1)/2 + 2n - 1$ independent parameters, one more than the minimal number required for flexibility. Because the BTL is a generalization of the log TL, it is clearly a flexible functional form: given any set of shares and elasticities consistent with a theoretically plausible demand system, we can impose the adding-up constraint (145) on the β's and select the parameters so as to obtain the log TL that reproduces the given shares and elasticities. This procedure, however, fails to exploit the BTL's advantage over the log TL: the presence of an additional parameter, B. The key to exploiting this advantage is the observation that the subset of BTL forms corresponding to any value of B is flexible: given an arbitrary value of B, we can choose the remaining BTL parameters to mimic the shares, the income elasticities, and the price elasticities of any theoretically plausible demand system at the point of approximation. Empirical applications of the BTL rest on this fact, since there is no reason to expect the estimated value of B to be 0.

To prove that the BTL is flexible for an arbitrary value of B, $\bar{B}$, we select the BTL parameters as follows: let the α's equal the shares at the point of approximation; let the β's be given by the values corresponding to (164) and (165) where ε is equal to $2\bar{B}$ and D is unity:

$$(176) \qquad \beta_{ii} = \bar{w}_i + \bar{w}_i \bar{E}_i^i + \bar{w}_i^2 - \bar{w}_i^2 \bar{E}_\mu^i + \bar{w}_i^2 \bar{B}$$

$$(177) \qquad \beta_{ij} = \bar{w}_i \bar{E}_j^i + \bar{w}_i \bar{w}_j - \bar{w}_i \bar{w}_j \bar{E}_\mu^j + \bar{w}_i \bar{w}_j \bar{B}.$$

We must show that the β's specified in this way satisfy the symmetry requirements, $\beta_{ij} = \beta_{ji}$, and that they satisfy the adding-up condition. Symmetry of the β's follows from the Slutsky symmetry conditions, most conveniently expressed in (152c). To show that the BTL adding-up condition holds, we sum (176) and (177) over all j and find

$$(178) \qquad \sum_j \beta_{ij} = \bar{w}_i - \bar{w}_i \bar{E}_\mu^i + \bar{w}_i \bar{B}$$

by an argument parallel to that used to derive (173). Summing (178) over i yields (144). It is straightforward to verify that the BTL with the parameters chosen in this way mimics the behavior of the original demand system at the point of approximation.

---

[9]These parameter counts refer to flexible functional forms for consumer demand systems. Counting parameters in producer demand system (i.e., demands for factors of production) is different because output is measurable while utility is not.

3.3. Advantages and Limitations

Flexible functional forms provide second order approximations to arbitrary twice differentiable functions. Instead of focusing on the ability of flexible functional forms to approximate preference orderings (i.e., to provide second order approximations to direct or indirect utility functions representing preferences), we have focused on their ability to provide first order approximations to theoretically plausible demand systems derived from preferences. The two approaches are fully consistent: since the partial derivatives of the demand functions are uniquely determined by the second partial derivatives of the utility function, the ability to approximate an arbitrary utility function implies the ability to approximate an arbitrary theoretically plausible demand system.[10]

We do not believe that the problem of modeling price effects has been solved by the development of flexible functional forms. Although the introduction of these forms constitutes an important expansion of the menu of specifications available for empirical research, they have yet to receive the critical scrutiny they require. In this section we discuss issues relevant to the assessment of flexible functional forms.

Suppose we want to estimate all of the elasticities of a demand system at a specified point of approximation in the price–expenditure space; suppose further that we can collect price–quantity data on which to base our estimates at any price–expenditure situations we select. Under these assumptions a plausible strategy is to specify a flexible functional form and estimate it using data corresponding to price–expenditure situations near the point of approximation. This strategy has seldom been implemented, presumably because economists seldom can select the price–expenditure situations at which data are collected.[11]

Viewed as local approximations, flexible functional forms avoid the restrictions inherent in nonflexible specifications; in doing so, however, they must either introduce additional parameters or imply other restrictions. For example, a nonflexible form with an equal number of parameters might be capable of approximating the second partial derivatives of the demand functions with respect to own-price and expenditure, at the cost of imposing restrictions on the cross-price derivatives. Thus, it is misleading to speak of a flexible functional form with K parameters as being "less restrictive" than a nonflexible specification with K independent parameters.

---

[10]To express the partial derivatives of the demand functions in terms of the second partials of the utility function, differentiate the first order conditions for utility maximization and solve using Cramer's rule, as in Samuelson [1947, pp. 100–102].

[11]There has been work based on experimental data, although little of it involves demand system estimation. The most prominent example of experimental data are those generated by time-of-day pricing experiments in electricity, by health care utilization experiments, and by negative income tax experiments.

Now suppose that we cannot select the price–expenditure situations at which data are collected, but instead must use price–quantity data corresponding to "natural" (i.e., uncontrolled, nonexperimental) price situations; for example, suppose our price data are deflators from the national product accounts or price indexes from the CPI. With such data the "local approximation" argument is irrelevant. To invoke it would require us to base our estimates on data from a neighborhood of the point of approximation and, without data from controlled experiments, the number of observations near the point of approximation is insufficient to permit estimation. To justify estimating a demand system using observations corresponding to price situations removed from the point of approximation, we must assume that the demand system we are estimating holds in a region large enough to include every data point we are using to estimate it. To motivate this assumption, it is often convenient to assume that it holds in a larger region, for example, in a connected region that includes the convex hull of the sample points. We call a demand system that holds in a region of the price–expenditure space an "exact" functional form in that region. Flexible functional forms, so appealing when viewed as local approximations, lose much of their appeal when interpreted as exact specifications.

When the number of goods is small, an attractive feature of flexible functional forms is that, compared to frequently estimated systems such as the LES and the generalized CES, they involve estimation of a large number of independent parameters. For example, with three goods the BTL involves eight parameters and the LES five. This fact, however, does not provide an argument for flexible functional forms because it applies equally to flexible and nonflexible functional forms involving the same number of parameters. For example, with three goods both the BTL and the QES contain eight parameters and it is not known whether the QES with three goods is flexible. Thus, although flexible functional forms have accustomed economists to estimating specifications involving many parameters, the availability of nonflexible multiparameter specifications precludes using this fact to justify the selection of a flexible functional form.

When the number of goods is large, the number of parameters in flexible functional forms, which increases with the square of the number of goods, becomes a liability. For example, with 15 goods the LES involves 29 parameters; the log TL, a parsimonious flexible functional form, involves 133. Empirical demand analysis has thus far avoided dealing with an unmanageable number of parameters by confining itself to systems involving a small number of goods.[12] Parsimonious or near parsimonious flexible functional forms are most attractive if one believes that the demand

---

[12]To retain at least some of the local approximation advantages of.flexible functional forms while avoiding this proliferation of parameters, Diewert and Wales [1988] define and estimate "semiflexible" functional forms.

for a particular good is influenced as much by the price of any randomly selected good as by its own price or by expenditure. That is, a parsimonious flexible functional form corresponds to a "flat" or "uninformative" prior; it is this prior that justifies an allocation of parameters in which each specific cross-price effect, each own-price effect, and each expenditure effect receives equal weight. Because we believe that in most situations a randomly chosen specific cross-price effect is likely to be less important than the own-price or expenditure effect, we are skeptical of the allocation of parameters corresponding to parsimonious flexible functional forms. Flexible and nonflexible functional forms impose different restrictions on demand behavior, but in the absence of a "flat" prior, flexible functional forms may well be as restrictive as nonflexible specifications containing the same number of parameters. We emphasize this point because when flexible functional forms were first introduced, excessive claims were made for them. Although we dispute the excessive portions of these claims, the space we have devoted to flexible functional forms reflects our assessment of their importance.

When the data offer extensive variation in prices and expenditure and degrees of freedom are not a problem, this objection can be met by the use of specifications that are flexible and that include additional parameters to allow more "flexibility" in capturing expenditure and own-price effects. The generalized translog (GTL) of Pollak and Wales [1980] is an example of a specification of this sort.[13] When the available data do not permit estimation of additional parameters, we recommend sacrificing flexibility in order to obtain specifications better adapted to capturing own-price and expenditure effects.

It is often useful to view the choice of a functional form as a two-stage process. In the first stage, the number of independent parameters in the functional form is chosen; in the second, a particular form with the specified number of parameters is selected. Decomposing the choice in this way draws attention to the problem of allocating a fixed number of parameters so as to capture best the price and expenditure responses we want to estimate from a particular data set. The nature of the data set itself will almost certainly affect the choice of a functional form: household budget data usually offer substantial variation in expenditure levels and limited variation in relative prices. Time series data, on the other hand, usually offer substantially more variation in relative prices and less variation in expenditure. Thus, household budget data offer greater scope for estimating expenditure effects and should be analyzed using functional forms capable of reflecting them. Time series data, on the other hand, are likely to offer greater opportunities than household budget data for

---

[13]In the production context, the CES-translog of Pollak, Sickles, and Wales [1984] is another example.

estimating the specific cross-price effects that flexible functional forms are designed to model.

APPENDIX A: THE GORMAN POLAR FORM THEOREM

Theorem: Any theoretically plausible demand system linear in expenditure

$$(A1) \qquad\qquad h^i(P, \mu) = C^i(P) + B^i(P)\mu$$

must be of the form

$$(A2) \qquad\qquad h^i(P, \mu) = f_i(P) - \frac{g_i(P)}{g(P)} f(P) + \frac{g_i(P)}{g(P)} \mu$$

where $f(P)$ and $g(P)$ are functions homogeneous of degree 1. These demand functions are generated by the indirect utility function

$$(A3) \qquad\qquad \psi(P, \mu) = \frac{\mu - f(P)}{g(P)}.$$

An obvious corollary is that any demand system exhibiting expenditure proportionality

$$(A4) \qquad\qquad h^i(P, \mu) = B^i(P)\mu$$

must be of the form

$$(A5) \qquad\qquad h^i(P, \mu) = \frac{g_i(P)}{g(P)} \mu$$

and is generated by the indirect utility function

$$(A6) \qquad\qquad \psi(P, \mu) = \frac{\mu}{g(P)}.$$

Lemma: If the demand system (A1) is theoretically plausible, then

$$(A7) \qquad\qquad \sum p_k B^k = 1$$

$$(A8) \qquad\qquad \sum p_k C^k = 0$$

$$(A9) \qquad\qquad B^i_j = B^j_i$$

$$(A10) \qquad\qquad C^i_j + C^j B^i = C^j_i + C^i B^j.$$

These equalities hold as identities in P.

Proof of Lemma: The budget constraint

$$\mu = \mu \sum p_k B^k + \sum p_k C^k$$

holds as an identity, so terms in like powers of μ must be equal. This implies

(A7) and (A8). The Slutsky symmetry conditions imply that

$$\frac{\partial h^i}{\partial p_j} + h^j \frac{\partial h^i}{\partial \mu}$$

is symmetric in i and j (i.e., is equal to the corresponding expressions with i and j interchanged) and these equalities hold identically in (P, μ). Calculating this expression from (A1) we find

$$\frac{\partial h^i}{\partial p_j} + h^j \frac{\partial h^i}{\partial \mu} = (B_j^i + B^j B^i)\mu + (C_j^i + C^j B^i).$$

Since this holds as an identity in μ, terms in like powers of μ are equal to the corresponding terms with i and j interchanged. Equating terms in like powers of μ and cancelling where possible yields (A9) and (A10).   QED

Proof of Theorem: The final assertion is easily verified using Roy's identity. The proof of the first part is broken down into two steps: first, we show that there exists a function $f(P)$, homogeneous of degree 1, such that (A1) can be written as

(A11) $h^i(P, \mu) = B^i(\mu - f) + f_i.$

Second, we show that there exists a function $g(P)$, homogeneous of degree 1, such that $g_i/g = B^i$. We assume throughout that the demand functions are differentiable enough to support the calculus arguments we employ and, hence, we establish our results only for this subclass.

In each part of the proof we appeal to a mathematical theorem on the existence of local solutions to systems of partial differential equations. The theorem guarantees the existence of a function $z = T(P)$, which satisfies a system of partial differential equations

$$\frac{\partial z}{\partial p_i} = \phi^i(P, z)$$

provided the symmetry conditions

$$\phi_j^i(P, z) + \phi_z^i(P, z)\phi^j(P, z) = \phi_i^j(P, z) + \phi_z^j(P, z)\phi^i(P, z)$$

hold (see Hurwicz and Uzawa [1971, Appendix]).

First, we show the existence of an f that satisfies

(A12) $C^i = -B^i f + f_i$

or, equivalently,

(A13) $f_i = C^i + B^i f.$

We define the functions $\phi^i(P, z)$ by

$$\phi^i(P, z) = C^i + B^i z.$$

By direct calculation

$$\phi^i_j + \phi^i_z \phi^j = C^i_j + B^i C^j + B^i_j z + B^i B^j z.$$

The fourth term is clearly symmetric. The symmetry of the first two terms is implied by (A10) and the symmetry of the third term by (A9). This establishes the existence of the function f(P).

To show that f(P) is homogeneous of degree 1, we multiply (A13) by p_i and sum over all goods to obtain

$$\sum p_k f_k = \sum p_k C^k + f \sum p_k B^k.$$

Making use of (A7) and (A8) we find

$$\sum p_k f_k = f.$$

The converse of Euler's theorem implies that f is homogeneous of degree 1.

Second, we show the existence of a function g(P) such that $g_i/g = B^i$. To do this, we define the function $\phi^i(P, z)$ by $\phi^i(P, z) = B^i(P)z$. Calculating $\phi^i_j + \phi^i_z \phi^j$, we find

$$\phi^i_j + \phi^i_z \phi^j = B^i_j z + B^i B^j z.$$

The second term is clearly symmetric and the first is symmetric in the light of (A9). Hence, there exists a function g(P) such that $g_i = B^i g$.

To show that g(P) is homogeneous of degree 1, we substitute g_i/g for B^i in (A7) and multiply by g to obtain

$$\sum p_k g_k = g.$$

The converse of Euler's theorem implies that g is homogeneous of degree 1.

QED

APPENDIX B: INTERPRETATION OF LOG TL AND BTL PARAMETERS

For the log TL, by differentiating the share equations (136) with respect to $\log p_i$, $\log p_j$, and $\log \mu$, we obtain equations analogous to (156), (157), and (159):

(B1)
$$w_i(1 + E^i_i) = \frac{\beta_{ii}}{D} - w_i \frac{\sum \beta_{ki}}{D}$$

(B2)
$$w_i E^i_j = \frac{\beta_{ij}}{D} - w_i \frac{\sum \beta_{kj}}{D}, \quad i \neq j,$$

(B3)
$$w_i E^i_\mu - w_i = - \frac{\sum_j \beta_{ij}}{D},$$

where D is the denominator of the log TL share equation

(B4)
$$D(P) = 1 + \sum\sum \beta_{kj} \log p_j.$$

Using the fact that the β's are symmetric, we substitute (B3) into (B1); solving for β_{ii}/D yields

(B5)
$$\frac{\beta_{ii}}{D} = w_i[(1 + E^i_i) - w_i(E^i_\mu - 1)].$$

Similarly, substituting (B3) into (B2) and solving for β_{ij} yields

(B6)
$$\frac{\beta_{ij}}{D} = w_i E^i_j + w_i w_j - w_i w_j E^j_\mu, \qquad i \neq j.$$

These expressions hold for all values of P and μ, but they simplify considerably when all prices and expenditure are unity because $D(1, \ldots, 1, 1) = 1$.

Differentiating the BTL share equations with respect to $\log p_i$ and $\log p_j$ yields expressions that look identical to (B1) and (B2), the expressions obtained from the log TL. They differ, however, in the role played by expenditure. In the log TL these expressions are independent of expenditure; in the BTL, expenditure enters through the function w_i in the numerator and the function D in the denominator:

(B7)
$$D(P, \mu) = 1 + \sum\sum \beta_{kj} \log p_j - B \log \mu.$$

When all prices and expenditure are unity, these expressions simplify because $D = 1$. In the BTL case we differentiate the share equations with respect to $\log \mu$, obtaining

(B8)
$$w_i E^i_\mu - w_i = -\frac{\sum_j \beta_{ij}}{D} + w_i \frac{B}{D}$$

where B is given by (144).

To clarify the meaning of the BTL parameters, we begin by differentiating (B8) with respect to $\log \mu$. After some rearranging this yields

(B9)
$$\frac{\partial(w_i E^i_\mu - w_i)/\partial \log \mu}{w_i E^i_\mu - w_i} = \frac{2B}{D}.$$

This expression is independent of the choice of i, and we denote its common value by ε:

(B10)
$$\varepsilon(P, \mu) = \frac{\partial(w_i E^i_\mu - w_i)/\partial \log \mu}{w_i E^i_\mu - w_i} = \frac{\partial^2 \omega^i(P, \mu)/\partial(\log \mu)^2}{\partial \omega^i(P, \mu)/\partial \log \mu} = \frac{2B}{D}.$$

In the case of the log TL, $B = 0$ and, hence, $\varepsilon = 0$. In the case of the BTL, these expressions are the same for all goods, and their common value is $2B/D$; when all prices and expenditure are unity, this reduces to $2B$.

To obtain a behavioral interpretation of the β parameters in the BTL, we begin by rewriting (B8) as

(B11)
$$\frac{\sum \beta_{ij}}{D} = w_i - w_i E^i_\mu + w_i \frac{B}{D} = w_i - w_i E^i_\mu + \tfrac{1}{2} w_i \varepsilon.$$

Relying on the symmetry of the β's and substituting this expression into (B1) and (B2) yields

(B12)
$$\frac{\beta_{ii}}{D} = w_i [(1 + E^i_i) - w_i(E^i_\mu - 1)] + \tfrac{1}{2} w_i^2 \varepsilon$$

(B13)
$$\frac{\beta_{ij}}{D} = w_i E^i_j + w_i w_j - w_i w_j E^j_\mu + \tfrac{1}{2} w_i w_j \varepsilon, \qquad i \neq j.$$

These are identical to Eqs. (164) and (165) in the text. As we noted there, these expressions simplify considerably when all prices and expenditure are unity, because in that case $D = 1$ and $\varepsilon = 2B$.

3

Demographic Specification

Demographic variables such as family size and age composition are major determinants of household consumption patterns.[1] In this chapter we describe five general procedures for incorporating demographic variables into theoretically plausible demand systems. The procedures are general in the sense that they do not assume that the original demand system has a particular functional form, but can be used in conjunction with any complete demand system. In Chapter 6 we use these procedures to incorporate a single demographic variable, the number of children in the family, into a "generalized CES" demand system and two demographic variables, the number and age of children, into the QES and GTL.

We conclude this chapter by discussing the problem of comparing the welfare of households with different demographic profiles. The paradigmatic question is: "How much would a family with three children have to spend to make it as well off as a family with two children spending $12,000?" Some would argue that this question is meaningless. We do not. We do argue, however, that even if this question is meaningful, it cannot be answered by comparing the consumption patterns of households with two children with those of households with three children. The difficulty is that, although observing the demand behavior of households with different demographic profiles reveals their conditional preferences (i.e., their preferences over vectors of goods taking the demographic profiles as fixed), it cannot reveal their unconditional preferences (i.e., their preferences over goods *and* demographic profiles). The expenditure level required to make a three-child family as well off as it would be with two children and $12,000 depends on how the family feels about children; observed differences in the consumption patterns of two- and three-child families cannot even tell us whether the third child is regarded as a blessing or a curse.

---

[1] This chapter draws on Pollak and Wales [1979, 1981, 1982] and Pollak [1991].

1. GENERAL PROCEDURES

This section discusses five general procedures for incorporating demographic variables into classes of demand systems and the related notion of "economies of scale in consumption." We begin with an original class of demand systems, $\{x_i = \bar{h}^i(P, \mu)\}$. We assume that these original demand systems are theoretically plausible, denoting the corresponding indirect utility function by $\bar{\psi}(P, \mu)$ and the direct utility function by $\bar{U}(X)$. Each procedure replaces this original class of demand systems by a related class involving additional parameters, and postulates that only these additional parameters depend on the demographic variables. The specification is completed by postulating a functional form relating these newly introduced parameters to the N demographic variables, $(\eta_1, \ldots, \eta_N)$. Lewbel [1985] shows that all of the procedures we describe that yield theoretically plausible demand systems are special cases of a "modifying function" procedure. Lewbel's procedure modifies the expenditure function—in effect, by first replacing each price by a function that depends on all prices and demographic variables and then subjecting the resulting expenditure function to a further transformation that depends on all prices and demographic variables.

The treatment of demographic effects in the context of theoretically plausible demand systems dates from Barten [1964]. Recent work in this vein includes Parks and Barten [1973], Lau, Lin, and Yotopoulos [1978], Muellbauer [1977], and Pollak and Wales [1978, 1980]. In this respect the classic work of Prais and Houthakker [1955] is the culmination of an earlier tradition that analyzed the demand for each good separately and without reference to utility maximization. Prais and Houthakker use expenditure as their independent variable and construct an exhaustive set of consumption categories. But they rely on a single equation rather than a complete system approach and most of their estimates are based on double-log and semi-log demand equations that, when summed over all goods, do not satisfy the budget identity. Prais and Houthakker [1955, pp. 83–85] recognize and discuss the "adding-up" problem but do not develop a satisfactory treatment. Maximum likelihood estimation of complete demand systems takes full account of the budget constraint in the demand equations and the stochastic structure.

The five procedures we consider are demographic translating (Section 1.1); demographic scaling (Section 1.2); the "Gorman procedure," a specification that includes both translating and scaling as special cases (Section 1.3); the "reverse Gorman procedure," a new procedure that is a mirror image of Gorman's procedure (Section 1.3); and a specification we call the "modified Prais–Houthakker procedure" (Section 1.4). Section 1.5 discusses "economies of scale in consumption."

1.1. Demographic Translating

Demographic translating replaces the original demand system by

$$(1) \qquad h^i(P, \mu) = d_i + \bar{h}^i(P, \mu - \sum p_k d_k)$$

where the d's are translation parameters that depend on the demographic variables, $d_i = D^i(\eta)$.[2] In some cases translating can be interpreted as allowing "necessary" or "subsistence" parameters of a demand system to depend on the demographic variables. In other cases, however, this interpretation is misleading. There are two distinct problems. First, the d's can be negative. Second, with some demand systems regularity conditions require d_i to be greater than x_i and, in these cases, it is possible to interpret the d's as "bliss points" but not as "necessary quantities"; the demand system generated by the additive quadratic direct utility function provides a familiar example (see Chapter 2). If the original demand system is theoretically plausible, then the modified system is also, at least when d is close to 0. The modified system satisfies the first order conditions corresponding to the indirect utility function $\psi(P, \mu) = \bar{\psi}(P, \mu - \sum p_k d_k)$ and the direct utility function $U(X) = \bar{U}(x_1 - d_1, \ldots, x_n - d_n)$.

When translating is used to introduce demographic characteristics into complete demand systems, there is a close relation between the effects of changes in demographic variables and the effects of changes in total expenditure. The total effect on $p_i h^i$ of a change in η_t is given by

$$(2) \qquad \frac{\partial p_i h^i}{\partial \eta_t} = p_i \frac{\partial D^i}{\partial \eta_t} - \frac{\partial p_i h^i}{\partial \mu} \sum p_k \frac{\partial D^k}{\partial \eta_t},$$

where $\partial p_i h^i / \partial \mu$ is the marginal budget share for the ith good. If demographic effects are incorporated by translating, then we can use this formula to decompose the total effect into a "specific effect," $p_i \partial D^i / \partial \eta_t$, and a "general effect," $(\partial p_i h^i / \partial \mu) \sum p_k (\partial D^k / \partial \eta_t)$. In Section 1.2 we describe an alternative decomposition that presupposes demographic scaling. Neither decomposition coincides with the Prais and Houthakker distinction between "specific scales" and "income scales," which we discuss in Section 1.4.

A change in η_t causes a reallocation of expenditure among the consumption categories, but since total expenditure remains unchanged, any increases in the consumption of some goods must be balanced by decreases in the consumption of others. The sign of the effect on $p_i h^i$ of a change in

---

[2]Demographic translating was first employed by Pollak and Wales [1978]. Gorman [1976] describes a more general procedure, which includes demographic translating and demographic scaling as special cases.

η_t cannot be inferred from the sign of its effect on d_i; changes in the demographic variables affect all of the d's simultaneously, and the specific effect, $p_i \partial D^i / \partial \eta_t$, may be outweighed by the general effect $(\partial p_i h^i / \partial \mu)(\sum p_k \partial D^k / \partial \eta_t)$. Furthermore, there is no a priori presumption that an increase in a demographic variable such as family size will increase rather than decrease d_i since changes in the d's, regardless of their direction, imply a reallocation of expenditure among the goods but leave total expenditure unchanged. In contrast, in a model of habit formation that we discuss in Chapter 4 a change in the past consumption of x_i influences a parameter analogous to d_i but leaves parameters analogous to the other d's unchanged. Hence, with habit formation the effect of a change in past consumption of a particular good can be inferred, because the "general effect" cannot outweigh the "specific effect" unless the marginal budget share is greater than unity.

With demographic translating the effects of changes in different demographic variables are closely related to one another. From (2) we see that $\partial p_i h^i / \partial \eta_t$ depends linearly on the n derivatives $\partial D^k / \partial \eta_t$; the coefficients of these derivatives are simply p_k times the marginal budget share of good i and these coefficients are the same regardless of which demographic variable we are considering. Since the term $\sum p_k \partial D^k / \partial \eta_t$ is independent of the good whose response we are considering, it is easy to establish a relationship between the effects of a change in a particular demographic variable on different goods, but the resulting equations are not particularly enlightening.

Linear demographic translating

$$(3) \qquad D^i(\eta) = \sum_{r=1}^{N} \delta_{ir} \eta_r$$

provides a convenient specification of the function relating the translating parameters to the demographic variables and is the one we estimate in Chapter 6. We have not included constant terms in the definition of linear demographic translating (as we unfortunately did in Pollak and Wales [1978]) because such constants are better treated as part of the specification of the original demand system than as part of the demographic specification. Regardless of whether constant terms are treated as part of the demographic specification or the original demand system, however, they should be included whenever demographic translating is used; any demand system that does not include "translation parameters" (i.e., constant terms) can easily be modified to incorporate them. If linear demographic translating is applied to a demand system that does not include translation parameters, and if the original system is misspecified (i.e., if the true system includes translation parameters), then a demographic variable that exhibits little variation may appear significant because it acts as a proxy for the omitted constant terms. With linear demographic translating (2) becomes

(4)
$$\frac{\partial p_i h^i}{\partial \eta_t} = p^i \delta_{it} - \frac{\partial p_i h^i}{\partial \mu} \sum p_k \delta_{kt}.$$

Linear demographic translating adds at most $n \times N$ independent parameters to the original demand system ("at most" because in some demand systems, such as the nonhomogeneous fixed-coefficient system, not all of the d's are identified).

1.2. Demographic Scaling

Demographic scaling replaces the original demand system by

(5)
$$h^i(P, \mu) = m_i \bar{h}^i(p_1 m_1, \ldots, p_n m_n, \mu)$$

where the m's are scaling parameters that depend on the demographic variables, $m_i = M^i(\eta)$. Demographic scaling was first proposed by Barten [1964]; see Muellbauer [1977] for a discussion of demographic scaling and an attempt to test it. If the original demand system is theoretically plausible, then so is the modified system, at least for m's close to 1. The modified system satisfies the first order conditions corresponding to the indirect utility function $\psi(P, \mu) = \bar{\psi}(p_1 m_1, \ldots, p_n m_n, \mu)$ and the direct utility function $U(X) = \bar{U}(x_1/m_1, \ldots, x_n/m_n)$.

If the scaling functions are the same for all goods, we can interpret their common value as reflecting the number of "equivalent adults" in the household. If the scaling functions differ from one good to another, then m_i measures the number of equivalent adults on a scale appropriate to good i. In either case we can view both preferences and demand behavior in terms of demographically scaled prices and quantities. That is, we can interpret the households' preferences as depending not on the number of gallons of milk consumed, but on gallons per (milk) equivalent adult. Similarly, the relevant price—the price corresponding to x_i/m_i—is not the price per gallon, but the price per gallon per (milk) equivalent adult, $p_i m_i$. The interpretation of the m's as commodity-specific "equivalent adults" should not be taken too literally. Indeed, there is not even a presumption that an increase in a demographic variable will increase rather than decrease the m's, since changes in the m's, regardless of their direction, imply a reallocation of expenditure among the goods but leave its total unchanged.

Under demographic scaling the effects of changes in demographic variables are closely related to the effects of price changes. The relationship is most clearly visible in elasticity form. Let E_k^i denote the elasticity of demand for the ith good with respect to the kth price and $\hat{M}_t^k$ the elasticity of m_k with respect to η_t. Then it is easy to verify by differentiating (5) that the elasticity of demand for the ith good with respect to η_t, E_t^i, is given by

(6)
$$E_t^i = \hat{M}_t^i + \sum E_k^i \hat{M}_t^k.$$

Somewhat inelegantly, but without loss of clarity, we denote the elasticity of the ith good with respect to the price of goods i, j, and k by E_i^i, E_j^i, and E_k^i, and the elasticity with respect to the demographic variables r, s, and t by E_r^i, E_s^i, and E_t^i. We use k as a summation index for goods and r as a summation index for the demographic variables. This provides a very simple way to obtain the results of Muellbauer [1974, p. 105] and Barten [1964, p. 282]. If demographic effects are incorporated by scaling then we can use (6) to decompose the impact on good i of a change in a demographic variable into a "specific" and a "general" effect: the specific effect is the elasticity of m_i with respect to the demographic variable, and the general effect is a weighted sum of the elasticities of all the m's with respect to the demographic variable, with the price elasticities as weights. This decomposition suggests that the sign of E_t^i, the effect of η_t on h^i, cannot be inferred from the sign of $\hat{M}_t^i$, its effect on m_i, since the general effect may outweigh the specific effect.

With demographic scaling, as with demographic translating, the effects of changes in different demographic variables are closely related. From (6) E_t^i is a weighted sum of $\hat{M}_t^k$ values, and the weights, which involve price elasticities, are the same for all demographic variables. It is only through these weights that prices and expenditures enter the demographic elasticities. It is possible to eliminate a cross-price elasticity between the equations for E_t^i and E_r^i, but the resulting expression is not particularly transparent.

Linear demographic scaling, the specification we estimate in Chapter 6, is given by

$$(7) \qquad M^i(\eta) = 1 + \sum_{r=1}^{N} \varepsilon_{ir}\eta_r.$$

If the ε's are independent of i, then the common value of the m's reflects the number of equivalent adults in the household. Linear demographic scaling adds at most n × N independent parameters to the original demand system. In certain exceptional cases not all of the n scaling parameters are identified. The demand system implied by the Cobb–Douglas utility function provides an extreme example in which none of the scaling parameters is identified. It might seem natural to include a multiplicative constant in the specification of linear demographic scaling:

$$(8) \qquad M^i(\eta) = m_i^*\left(1 + \sum_{r=1}^{N} \varepsilon_{ir}\eta_r\right).$$

We have not done so for two reasons. First, such a constant term is better described as part of the specification of the original demand system than as part of the demographic specification. Second, the m*'s are not identified in any class of demand systems that is "closed under unit scaling." (A demand system is closed under unit scaling if, whenever some demand system in the class is exactly consistent with the data in one set of units, and the data are rescaled in new units, there exists another demand system in the class that is exactly consistent with the rescaled data.) We believe that only closed classes of demand systems should be used for empirical demand analysis. Any class not closed under unit scaling can easily be generalized to one that is. Virtually all classes of demand systems that have been used in demand analysis are closed under unit scaling; Diewert's "Generalized Cobb–Douglas" appears to be the only exception (see Wales and Woodland [1976, footnote 10] and Berndt, Darrough, and Diewert [1977, footnote 11]).

With "log linear demographic scaling"

$$(9) \qquad\qquad M^i(\eta) = \prod_{r=1}^{N} (\eta_r)^{\varepsilon_{ir}}$$

the ε's are the elasticities of the m's with respect to the demographic variables and (6) becomes

$$(10) \qquad\qquad E_t^i = \varepsilon_{it} + \sum E_k^i \varepsilon_{kt}.$$

1.3. Gorman Specifications

Gorman [1976] proposed a specification that replaces the original demand system by

$$(11) \qquad h^i(P, \mu) = d_i + m_i \bar{h}^i(p_1 m_1, \ldots, p_n m_n, \mu - \sum p_k d_k)$$

where the d's and the m's (and only these parameters) depend on the demographic variables. This demand system is theoretically plausible (at least for d's near 0 and m's near 1) and is generated by the indirect utility function $\psi(P, \mu) = \bar{\psi}(p_1 m_1, \ldots, p_n m_n, \mu - \sum p_k d_k)$ and the direct utility function $U(X) = \bar{U}[(x_1 - d_1)/m_1, \ldots, (x_n - d_n)/m_n]$. Demographic translating corresponds to the special case in which the m's are unity, and demographic scaling corresponds to the case in which the d's are 0. Gorman's specification can be obtained from the original demand system by first scaling and then translating.

The "reverse Gorman" specification is obtained by first translating and then scaling. This yields the demand system

$$(12) \qquad h^i(P, \mu) = m_i[d_i + \bar{h}^i(p_1 m_1, \ldots, p_n m_n, \mu - \sum p_k m_k d_k)],$$

which satisfies the first order conditions corresponding to the indirect utility function $\psi(P, \mu) = \bar{\psi}(p_1 m_1, \ldots, p_n m_n, \mu - \sum p_k m_k d_k)$ and the direct utility function $U(X) = \bar{U}[(x_1/m_1) - d_1, \ldots, (x_n/m_n) - d_n]$.

If we impose particular forms for $D(\eta)$ and $M(\eta)$, then the Gorman and reverse Gorman procedures are distinct and the reverse Gorman procedure is new. For example, suppose the original demand system is linear in expenditure and we require linear forms for $D(\eta)$ and $M(\eta)$; then the reverse Gorman specification yields demand functions quadratic in the demographic variables, while the Gorman specification implies demand functions linear in them. If we do not impose particular forms, then the procedures are not distinct. This follows immediately from the observation that when we define new variables d_i by $d_i = d_i m_i$ and new functions $D^i(\eta)$ by $D^i(\eta) = D^i(\eta)M^i(\eta)$, substituting the new variables into the reverse Gorman specification yields the Gorman specification.

Postulating that the d's and m's are linear functions of the demographic variables allows us to test linear translating and linear scaling against more general hypotheses. The specification we estimate in Chapter 6

$$(13) \qquad\qquad D^i(\eta) = v \sum_{r=1}^{N} \beta_{ir}\eta_r$$

and

$$(14) \qquad\qquad M^i(\eta) = 1 + (1 - v) \sum_{r=1}^{N} \beta_{ir}\eta_r$$

requires only one more parameter than either linear translating or linear scaling separately; translating corresponds to $v = 1$, and scaling to $v = 0$. Replacing v by v_i in (13) and (14) yields a more general specification, which allows the balance between translating and scaling to differ for different goods; with one demographic variable, allowing the v_i's to differ is equivalent to assuming $D(\eta)$ and $M(\eta)$ are unrestricted linear forms given by (3) and (7).

1.4. The Modified Prais–Houthakker Procedure

The "modified Prais–Houthakker procedure" replaces the original demand system by

$$(15) \qquad\qquad h^i(P, \mu) = s_i \bar{h}^i(P, \mu/s_o)$$

where s_i is a "specific scale" for the ith commodity and depends on the demographic variables, $s_i = S^i(\eta)$, and s_o is an "income scale" implicitly defined by the budget constraint

$$(16) \qquad\qquad \sum p_k s_k \bar{h}^k(P, \mu/s_o) = \mu.$$

Thus, the income scale is a function of all prices and expenditure as well as the demographic variables: $s_o = S^0(P, \mu, s_1, \ldots, s_n)$. If the left-hand side of (16) is an increasing function of μ/s_o, then it defines s_o uniquely; the absence of inferior goods is sufficient to guarantee this.

Prais and Houthakker [1955, Chapter 9] proposed a technique for incorporating demographic variables into demand equations using a single income scale and a specific scale for each good, but they never reconciled their technique with an overall budget constraint. They used data from a single budget study to estimate the effects of changes in demographic variables and expenditure (but not prices) on household consumption patterns. Their proposal should be viewed against the background of earlier techniques.[3] Given a body of household budget data, the simplest way to introduce a single demographic variable such as family size is to postulate that per capita consumption of each good is a function of per capita expenditure. This per capita approach treats all types of individuals (e.g., adults and children) as identical and does not allow for systematic differences in their "needs" or "requirements." This defect can be remedied by postulating an "equivalence scale" that converts the household's demographic profile, η, into the corresponding number of "equivalent adults" and makes consumption per "equivalent adult" a function of expenditure per equivalent adult. This implies expenditure equations (with prices suppressed) of the form $p_i x_i / s = \phi^i(\mu/s)$ or $p_i x_i = s\phi^i(\mu/s)$ where s is given by $s = S(\eta)$.[4] In suppressing prices we have followed Prais and Houthakker and the traditional equivalence scale literature; since its focus was on the analysis of data from a single budget study, ignoring prices was relatively harmless. Barten [1964] was the first to emphasize the role of prices in connection with demographic variables, but he was primarily interested in using variations in demographic characteristics instead of variations in prices to identify all of the parameters of a theoretically plausible demand system. Muellbauer [1974] argues against Barten's contention that demographic scaling permits us to dispense with price variation; in any event, Barten's use of prices in conjunction with demographic variables in a complete demand system framework is an important development. But this specification assumes a single equivalence scale for all goods, the same for food as for housing, for milk as for beer. The Prais–Houthakker procedure avoids this assumption by introducing separate "specific scales" for each good. In particular, it postulates that the expenditure equations can be written as $p_i x_i = s_i \phi^i(\mu/s_0)$.[5]

Muellbauer [1977, 1980] proposed defining the income scale implicitly through the budget constraint.[6] If the specific scales are identical for all

---

[3] Prais and Houthakker provide references to this literature and note (p. 126) that their view is "essentially that expressed by Sydenstricker and King" [1921].

[4] For example, adult males = 1, adult females = 0.9, children under two = 0.1.

[5] Prais and Houthakker [1955, p. 128] assume a linear form for the specific scales. They are less explicit about the way the income scale depends on the demographic variables, but they do assume that it is independent of expenditure as is clear from their calculation of the income (i.e., expenditure) elasticity [1955, p. 129]. Barten [1964, pp. 283–286] also interprets the Prais–Houthakker procedure as requiring the income scale to be independent of expenditure.

[6] Muellbauer advances his proposal in the context of demand systems that rule out substitution in response to relative price changes, and argues that this restriction is needed to permit welfare comparisons between households with different demographic profiles. We reject his welfare comparisons even in the no-substitution case, for reasons discussed in Section 3. Regardless of one's views on welfare comparisons, the no-substitution restriction is unnecessary for analyzing the effects of demographic differences on consumption patterns, so that Muellbauer's proposal can be used to define empirically interesting demand specifications.

goods, then (16) implies that the income scale is also equal to this common value, which we can interpret as the number of "equivalent adults" in the household. Furthermore, with identical specific scales for all goods, the modified Prais–Houthakker procedure is equivalent to scaling with identical scaling functions for all goods. (To prove this, note that the demand functions are homogeneous of degree 0 in prices and expenditure; hence, the income scale can be viewed as multiplying all prices instead of dividing expenditure.)

The modified Prais–Houthakker procedure, unlike the other procedures we have considered, need not yield a theoretically plausible demand system. Indeed, in Pollak and Wales [1981] we showed that it does so only in very special cases, and conjectured that it does so if and only if the original demand system corresponds to an additive direct utility function. Lewbel [1986] proved that this conjecture is correct. Two theorems (whose proofs are given in Appendix A) lend credence to this conjecture.

Theorem 1: If the modified Prais–Houthakker procedure is applied to a demand system corresponding to an additive direct utility function

$$(17) \qquad \bar{U}(X) = \sum \bar{u}^k(x_k)$$

then the resulting demand system is theoretically plausible and corresponds to the additive direct utility function

$$(18) \qquad U(X) = \sum s_k \bar{u}^k(x_k/s_k).$$

Theorem 2: If the modified Prais–Houthakker procedure is applied to a theoretically plausible demand system linear in expenditure, then the resulting demand system is theoretically plausible only if the original demand system corresponds to an additive direct utility function.[7]

We illustrate the modified Prais–Houthakker procedure using an original demand system linear in expenditure, since in this case we can solve the budget identity to obtain an explicit expression for the income scale and a closed-form expression of the demand equation. We make use of the fact that any theoretically plausible demand system linear in expenditure can be written in the form

$$(19) \qquad \bar{h}^i(P, \mu) = f_i - \frac{g_i}{g} f + \frac{g_i}{g} \mu = f_i + \gamma^i(\mu - f)$$

where $\gamma^i = g_i/g$ and $f(P)$ and $g(P)$ are functions homogeneous of degree 1 (see Chapter 2). Using the modified Prais–Houthakker procedure, we

---

[7] The proof also assumes that $n \geqslant 3$, that no good has a 0 marginal budget share, and that the utility and demand functions are differentiable enough to support the calculus arguments employed in Pollak [1976b]. If $n = 2$, then the mathematical integrability conditions are automatically satisfied.

write the demand functions as

$$(20) \qquad h^i(P, \mu) = s_i f_i + s_i \gamma^i \left(\frac{\mu}{s_0} - f \right)$$

where s_0 is defined by

$$(21) \qquad \sum p_k s_k f_k + \sum p_k s_k \gamma^k \left(\frac{\mu}{s_0} - f \right) = \mu.$$

To find the implied demand functions, we solve (21) for $(\mu/s_0) - f$ and substitute into (20) to obtain

$$(22) \qquad h^i(P, \mu) = s_i f_i + \frac{s_i \gamma^i}{\sum p_k s_k \gamma^k} (\mu - \sum p_k s_k f_k).$$

We can solve (21) explicitly for the income scale

$$(23) \qquad S(P, \mu, s_1, \dots, s_n) = \frac{\mu \sum p_k s_k \gamma^k}{\mu - \sum p_k s_k f_k + f \sum p_k s_k \gamma^k},$$

which clearly depends on prices and expenditures as well as on the specific scales. If the original demand system exhibits expenditure proportionality, $\bar{h}^i(P, \mu) = \gamma^i(P)\mu$, then the income scale is independent of expenditure.

With the modified Prais–Houthakker procedure, the effects of changes in the demographic variables are closely related to the effects of changes in expenditure. The relationship can be stated most clearly in elasticity form, where E_o^i and $\hat{S}_o^0$ denote the expenditure elasticities of h^i and S^0, and $\hat{S}_t^i$ and $\hat{S}_t^0$ the elasticities of S^i and S^0 with respect to the demographic variable η_t. Taking the logarithmic derivatives of (15) with respect to μ and η_t, it is easy to verify that

$$(24) \qquad E_t^i = \hat{S}_t^i - \frac{E_o^i \hat{S}_t^0}{1 - \hat{S}_o^0}.$$

If demographic effects are incorporated by the modified Prais–Houthakker procedure, then we can use this formula to decompose the effect of a change in a demographic variable into a "specific effect" and a "general effect." The specific effect is the elasticity of s_i with respect to the demographic variable; the general effect is the product of the elasticity of s_0 and h^i. It is tempting to associate the decomposition with the Prais and Houthakker [1955, p. 129] distinction between "specific scales" and "income scales," but because the modified Prais–Houthakker procedure is not identical to the Prais–Houthakker procedure, the association is rather loose. This decomposition implies that the sign of E_t^i cannot be inferred from a knowledge of $\hat{S}_t^i$ alone, since the general effect need not operate in the same direction as the specific effect and may outweigh it.

There are some obvious cases in which inferences are possible. For example, if the income elasticity is 0, then $E_t^i = \hat{S}_t^i$. If the signs of $\hat{S}_t^i, \hat{S}_t^0, \hat{S}_o^0$, and E_o^i are known, then we can determine the sign of E_t^i in those cases in which the specific effect and the general effect work in the same direction.

The linear modified Prais–Houthakker specification, which we estimate in Chapter 6, is given by

$$(25) \qquad S^i(\eta) = 1 + \sum_{r=1}^{N} \sigma_{ir}\eta_r.$$

It adds at most $n \times N$ independent parameters to the original demand system. (As with translating and scaling, we have not included constant terms in our specification of the linear modified Prais–Houthakker procedure because we prefer to treat such constants as part of the original demand system.) To illustrate the "at most" assertion, suppose the original demand system corresponds to a Cobb–Douglas utility function, $\bar{h}^i(P, \mu) = (a_i / \sum a_k)(\mu/p_i)$; in this case, the modified Prais–Houthakker procedure yields $h^i(P, \mu) = (a_i s_i / \sum a_k s_k)(\mu/p_i)$. In the original demand system without demographic variables, only $n - 1$ of the a's are independently identifiable parameters. When demographic variables are introduced using the linear modified Prais–Houthakker procedure, only $N \times (n - 1)$ of the additional parameters are identifiable. The situation is similar for other demand systems exhibiting expenditure proportionality and corresponding to additive direct utility functions: the CES and the Leontief or homogeneous fixed coefficient system. If the σ's are independent of i, then the scales are the same for all goods and the modified Prais–Houthakker procedure is equivalent to scaling. In the "log linear modified Prais–Houthakker procedure"

$$(26) \qquad S^i(\eta) = \prod_{r=1}^{N} (\eta_r)^{\sigma_{ir}}$$

the σ's are the elasticities of the s's with respect to demographic variables and (24) becomes

$$(27) \qquad E_t^i = \sigma_{it} - \frac{E_o^i \hat{S}_t^0}{1 - \hat{S}_o^0}.$$

1.5. Economies of Scale in Consumption

"Economies of scale in consumption" is the rubric of empirical demand analysis under which investigators explore alternative functional forms relating demographic variables to parameters of demand systems. A different notion of "economies of scale in consumption" involves welfare comparisons between households with different demographic profiles; that notion is defined in terms of the effect of an increase in family size on the expenditure required to attain a particular standard of living. Although Prais and Houthakker [1955, Chapter 10] use welfare considerations to motivate their discussion of economies of scale in consumption, that discussion serves as a springboard for introducing more general functional forms relating the demographic variables to the specific scale and income scale.

Two notions of economies of scale in consumption do not involve demographic effects. The first presupposes a household production model (Becker [1965], Michael and Becker [1973], Pollak and Wachter [1975]) in which the household uses inputs of "market goods" and time to produce "basic commodities," which are the arguments of its utility function; economies of scale in consumption are then defined in terms of the returns to scale properties of the household's technology. The second notion assumes that the prices paid for the market goods are functions of the amounts purchased and defines economies of scale in consumption as the availability of "quantity discounts" for bulk purchases. These two notions are easily confused because economies of scale in the household technology may imply that the shadow prices of commodities fall as their consumption increases. Both of these nondemographic notions of economies of scale in consumption are discussed in David [1962, pp. 10–19].

In this section we discuss economies of scale in consumption in terms of demographic variables, introducing a quadratic generalization of the linear specification thus far considered. Demographic translating places no restrictions on the form of the functions $\{D^i(\eta)\}$, but in Section 1.1 we emphasized linear demographic translating, (3). One straightforward and tractable generalization of linear demographic translating is the quadratic specification,

$$(28) \qquad D^i(\eta) = \sum_{r=1}^{N} \delta_{ir}\eta_r + \sum_{r=1}^{N} \theta_{ir}\eta_r^2.$$

The quadratic specification does not require all demographic characteristics to have the same "economies of scale" implications, but this generality is achieved at the cost of introducing $n \times N$ more parameters than the linear specification.

With demographic scaling we considered both the linear and the log linear specifications. Log linear demographic scaling, (9), already permits economies of scale in consumption; we can restrict it and impose constant returns to scale by requiring the coefficients in (9) to satisfy

$$(29) \qquad \sum_{r=1}^{N} \varepsilon_{ir} = 1, \qquad i = 1,\ldots,n.$$

One simple generalization of linear demographic scaling, (7), is· the quadratic specification

$$(30) \qquad M^i(\eta) = 1 + \sum_{r=1}^{N} \varepsilon_{ir} \eta_r + \sum_{r=1}^{N} \theta_{ir} \eta_r^2$$

involving $n \times N$ additional parameters. These scaling specifications apply equally to the modified Prais–Houthakker procedure. In Chapter 6 we discuss estimates of demographic scaling and the modified Prais–Houthakker procedure based on this quadratic specification.

2. WELFARE COMPARISONS, SITUATION COMPARISONS, AND EQUIVALENCE SCALES

Economists find welfare comparisons troubling. Most of us are puzzled by questions such as, "How much would a woman with three children need to be as well off as a woman with two children spending $20,000?" or "How much would a family consisting of a woman and three children need to be as well off as a family consisting of a man, a woman, and three children spending $30,000?" Indeed, most economists think interpersonal welfare comparisons are nonsense.

Among economists who are not prepared to dismiss welfare comparisons out of hand, there is no consensus on what such comparisons mean or on what information is required to make them. Thus, if you ask an economist how much individual A needs to be as well off as individual B, the reply is more likely to challenge the question than answer it. Posed in terms of household A and household B, the welfare comparison question becomes even more tangled. Referring to families or households as "economic agents" not only fails to solve this problem, it also sweeps it under the rug. Recent work on bargaining models of marriage suggests that the notion of family or household preferences is problematic at best and probably untenable.

In this section we distinguish "welfare" comparisons from what we call "situation" comparisons. Situation comparisons compare alternative "price–demographic" situations on the basis of a single preference ordering. Exploiting the theory of the cost-of-living index, which provides the standard framework for comparing alternative price situations, we define a "generalized cost-of-living index" and argue that it provides a theoretical framework suitable for comparing alternative price–demographic situations.

Situation comparisons, which are based on the generalized cost-of-living index, require knowing only a single indifference map. Thus, the information required to make situation comparisons is easy to specify in theory. It is, however, difficult to obtain in practice. The difficulty is that the preferences needed to compare alternative price–demographic situations are "unconditional preferences" (in this case, preferences over

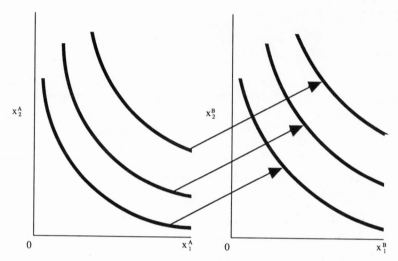

Figure 1   A correspondence between two indifference maps

price–demographic situations) and these preferences cannot be obtained by analyzing the consumption patterns of households with different demographic profiles.

Welfare comparisons require knowing two indifference maps and the "welfare correspondence" between them. Figure 1 illustrates a correspondence between two indifference maps. A welfare correspondence specifies which indifference curve on one individual's map yields the same welfare level as a particular curve on the other's map. Having found the corresponding curve, to answer the welfare comparison question we simply calculate the minimum expenditure required to attain that curve. Thus, like situation comparisons, welfare comparisons between individuals with different demographic profiles require first solving the practical problem of identifying their unconditional preferences. Unlike situation comparisons, welfare comparisons then require solving the additional theoretical and practical problems of making welfare comparisons between the indifference maps of two individuals.

In the absence of a theory of welfare comparisons, discussions of the practical problems of making such comparisons are pointless. Economists such as Deaton and Muellbauer [1980b, 1986] and Jorgenson [1990], who advocate specific empirical procedures for making interpersonal comparisons therefore must bear a double burden: first, establishing the general theoretical validity of making such comparisons and second, demonstrating the specific empirical validity of the particular procedures they advocate for making interpersonal welfare comparisons.

We begin with situation comparisons, because they are a prerequisite to welfare comparisons. We then discuss economists' changing views of welfare comparisons and alternative procedures for establishing a welfare correspondence between indifference maps. Finally, we discuss the complications inherent in the move from comparisons involving individuals to comparisons involving families or households.

2.1. Situation Comparisons

The cost-of-living index compares two price situations, evaluating them in terms of a single base preference ordering. More precisely, the cost-of-living index compares the expenditure required to attain a particular base indifference curve—the indifference curve of X^0—in the price situation P^a with that required to attain the same indifference curve in the price situation P^b.

The theory of the cost-of-living index, suitably generalized, provides a theoretical framework for making "situation comparisons," for example, comparisons of alternative "price–demographic" situations. Situation comparisons are based on a single preference ordering and do not require interpersonal comparisons.

The prerequisites for situation comparisons from the theory of the cost-of-living index can be summarized in two paragraphs.[8] Let X denote a vector of market goods and $R*$ a preference ordering over such vectors. The statement $X^a R* X^b$ means that X^a is at least as good as X^b evaluated in terms of the preference ordering $R*$. Corresponding to the preference ordering $R*$ is an "expenditure function," $E[P, X°; R*]$, showing the minimum expenditure required by an individual with preferences $R*$ to attain the indifference curve of $X°$ when facing prices P. The cost-of-living index, $I[P^a, P^b, X°; R*]$, is defined as the ratio of the minimum expenditure required to attain the base indifference curve at prices P^a to that required at prices P^b:

$$(31) \qquad I[P^a, P^b, X°; R*] = E[P^a, X°; R*]/E[P^b, X°; R*].$$

To construct a cost-of-living index, we first select a base preference ordering, $R*$, and then select from that preference ordering a base indifference curve, $X°$. The formal theory offers no guidance in selecting either the base preference ordering or the base indifference curve. The problem of selecting a base preference ordering arises in both spatial and intertemporal comparisons (see Pollak [1989, pp. 7–8]). Suppose, for example, we want to compare French and Japanese prices. If French and Japanese preferences differ, then constructing a cost-of-living index clearly requires the investigator to specify a base preference ordering. If French and Japanese preferences are identical, then it seems appropriate, at least in many cases, to base the comparison on this common preference ordering.

---

[8]Diewert [1981, 1983] and Pollak [1989] provide detailed discussions of the theory of the cost-of-living index.

Even if French and Japanese preferences are identical, however, another base preference ordering might be selected because it is more appropriate. For example, if the U.S. government wants to compare prices in Paris with those in Tokyo establish appropriate salary differentials for its diplomats, then U.S. preferences are presumably appropriate. In intertemporal comparisons, if preferences are constant over time, the selection of a base preference ordering is often implicit and invisible; if preferences change over time, however, the selection of a base preference ordering becomes explicit and evident.

Although the traditional cost-of-living index compares alternative situations that differ only in prices, it is straightforward to define a generalized cost-of-living index suitable for comparing situations that differ in other dimensions as well.[9] Formally, let Z denote a vector of "nonmarket goods"; these might include environmental variables (e.g., climate, air quality), public services provided by the government, and demographic variables. Let R denote a preference ordering over the consumption vectors (X, Z), where the statement $(X^a, Z^a) R (X^b, Z^b)$ is interpreted in the obvious way. The expenditure function corresponding to the preference ordering $R, E[(P, Z), (X^0, Z^0); R]$, shows the minimum expenditure on market goods required by an individual with preferences R facing prices P with nonmarket goods Z to attain the indifference curve of (X^0, Z^0). The generalized cost-of-living index compares the expenditure required to attain a particular base indifference curve—the indifference curve of (X^0, Z^0)—in the situation (P^a, Z^a) with that required to attain the same base indifference curve in the situation (P^b, Z^b). Formally, the generalized cost-of-living index, $I[(P^a, Z^a), (P^b, Z^b), (X^0, Z^0); R]$, is the ratio of these minimum expenditures:

$$(32) \qquad I[(P^a, Z^a), (P^b, Z^b), (X^0, Z^0); R]$$
$$= E[(P^a, Z^a), (X^0, Z^0); R]/E[(P^b, Z^b), (X^0, Z^0); R].$$

When nonmarket goods are identical in the two situations (i.e., $Z^a = Z^b = Z^*$), then the generalized cost-of-living index reduces to the conventional cost-of-living index for comparing alternative price situations. More precisely, it is the cost-of-living index corresponding to the "conditional" preference ordering over market goods, $R(Z^*)$, defined in the obvious way by $X^a R(Z^*) X^b$ if and only if $(X^a, Z^*) R (X^b, Z^*)$.

The distinction between conditional and unconditional preferences, first drawn in the context of the comparison problem by Pollak and Wales [1979], is crucial. Conditional preferences are defined over market goods, with nonmarket goods held fixed at specified levels. Thus, conditional preferences resemble conditional probabilities, which are "conditioned"

---

[9] The "generalized cost-of-living index" is developed in Pollak [1974; published, 1989], where it is called the "conditional environment-dependent cost-of-living index," and in Gillingham and Reece [1979, 1980].

on some specified event. Unconditional preferences, which are defined over the space of market goods and nonmarket goods, are analogous to unconditional or joint probabilities.

Making situation comparisons requires knowing unconditional preferences; conditional preferences do not contain enough information to compare situations that differ in the levels of the nonmarket goods.[10] An example illustrates the point: suppose an individual's preference ordering over market goods and nonmarket goods can be represented by a Cobb–Douglas utility function:

$$(33) \qquad U(X, Z) = \sum \alpha_k \log x_k + \sum \gamma_k \log z_k.$$

In this case, the individual's generalized cost-of-living index depends on the coefficients of both the market and nonmarket goods, but the individual's conditional preference ordering (i.e., preferences over market goods) is independent of the levels of the nonmarket goods. The problem of inferring unconditional preferences from conditional preferences is analogous to the problem of inferring unconditional probability distributions from conditional distributions; the hopelessness of the tasks is most apparent when preferences are separable or, analogously, when the probabilities are independent.

Treating the demographic variables as nonmarket goods, we can use the generalized cost-of-living index to compare price–demographic situations. That is, let η denote a "demographic profile" consisting of a single demographic variable such as family size or a vector of demographic variables such as race, sex, place of birth, and the ages and sexes of children. The unconditional preference ordering, R, enables us to make statements of the form $(X^a, \eta^a) R (X^b, \eta^b)$—that is, the situation (X^a, η^a) is at least as good as the situation (X^b, η^b).

The generalized cost-of-living index applies directly to comparisons involving demographic variables. The index, $I[(P^a, \eta^a), (P^b, \eta^b), (X^0, \eta^0); R]$, compares the expenditure required to attain a particular base indifference curve—the indifference curve of (X^0, η^0)—in the situation (P^a, η^a) with that required in the situation (P^b, η^b):

$$(34)$$
$$I[(P^a, \eta^a), (P^b, \eta^b), (X^0, \eta^0); R] = E[(P^a, \eta^a), (X^0, \eta^0); R] / E[(P^b, \eta^b), (X^0, \eta^0); R].$$

Although the term "situation comparison" is new, the idea is not. The generalized cost-of-living index is essentially what Pollak and Wales [1979] call an "unconditional equivalence scale." We now believe that the term "equivalence scale" causes so much confusion that it should be dropped from discussions of both situation and welfare comparisons and used only in the context of demographic variables in demand analysis.

We can, at least in principle, recover the unconditional preferences

---

[10] Pollak and Wales [1979] and Lewbel [1989] contain clear discussions of this point.

required to compare alternative price–demographic situations from unconditional choices. The usual revealed preference reasoning applies: if we observe an individual facing a choice between (X^a, η^a) and (X^b, η^b), and observe that the individual chooses (X^a, η^a), then we infer that (X^a, η^a) is at least as good as (X^b, η^b). As Pollak and Wales [1979] point out, this implies that "if a family chooses to have three children and $12,000 when it could have had two children and $12,000, then a revealed preference argument implies that the family prefers the alternative it chose." This link between unconditional preferences and unconditional choices suggests a potential problem with any "equivalence scale" implying that, at equal levels of expenditure, smaller families are better off than larger families. In the absence of government subsidization of families with children (e.g., through the tax system or through child allowances), if smaller families were better off and if contraception were free and perfectly effective, then fertility would be zero. Of course, subsidization of families with children might reflect society's concern with the present or future welfare of the children. These concerns, however, require moving from a conceptual framework that focuses exclusively on "individual" preferences to one that recognizes "households" or "families."

Unconditional preferences cannot be recovered from conditional choices. In terms of our Cobb–Douglas example, observing conditional choices enables us to recover the coefficients of the x's (up to a suitable normalization rule), but provides no information about the coefficients of the z's. Indeed, the observed conditional choices are compatible with any values—positive or negative—of the coefficients of the nonmarket goods. Even if we observe such an individual's choices of market goods for all possible budget constraints and all possible levels of the nonmarket goods, unless the individual chooses the levels of the nonmarket goods, we cannot infer how the individual would trade off environmental variables or public services against each other or against market goods.

Demographic variables present the same problem: unconditional preferences cannot be inferred from the consumption patterns of households that treat their demographic profiles as fixed when making consumption choices. Thus, as Pollak and Wales [1979] point out, if storks distributed babies randomly among households, then we could not even infer from the consumption patterns of two- and three-child households whether they regarded the third child as a blessing or a curse.

The Cobb–Douglas example also illustrates the everpresent possibility of identifying the coefficients of nonmarket goods or demographic variables by assumption when we have only data on conditional choices. Without fear of contradiction—the data, after all, reveal nothing about the coefficients of the z's—we can assume, for example, that the coefficients of all the z's are equal to a common value and set this common value to 0 (or to 1, or to −1). The possibility of observing unconditional fertility choices can thus be viewed either as an opportunity for empirical research or as a constraint on our ability to assume whatever we like. Conversely, given the impossibility of observing unconditional choices of race or sex, the fact that our assumptions about preferences over these variables are not rejected by the data provides cold comfort.

2.2. Welfare Comparisons

Interpersonal comparisons were readily accepted by late nineteenth-century economists such as Edgeworth for whom, to quote Samuelson [1947, p. 206], utility "was as real as his morning jam." Edgeworth and his fellow utilitarians thought that the statement "individual A is better off than individual B" had the same kind of empirical content as "individual A is heavier than individual B," or "individual A is taller than individual B." Utilitarian views remained dominant among economists until the 1930s, when the "ordinalist revolution" overthrew the twin assumptions that utility was cardinally measurable and interpersonally comparable. The ordinal utility or indifference curve approach of R. G. D. Allen, A. L. Bowley, and J. R. Hicks convinced economists that demand analysis did not require cardinal utility, and the "new welfare economics" of Lionel Robbins, with its focus on Pareto optimality, convinced economists that interpersonal comparisons had no scientific basis and, therefore, no place in "economic science."[11] By the 1940s economists had come to view interpersonal welfare comparisons with skepticism and suspicion. The neoclassical position, articulated by Samuelson in *Foundations of Economic Analysis* [1947], was that interpersonal comparisons (i) are unnecessary for demand analysis, (ii) cannot be made on the basis of observable demand behavior, and (iii) are, in the positivistic language of *Foundations*, "meaningless" in the technical sense that they have no refutable implications and, hence, no empirical content.

The interpretation of "observable behavior" is crucial to understanding Samuelson's claim that interpersonal comparisons are "meaningless." Samuelson regards hypotheses about preferences as meaningful only if they place refutable restrictions on choices from budget sets, and the revealed preference literature has generally followed his lead (see Pollak [1990]). Samuelson's contention that interpersonal comparisons are "meaningless" is an assertion that such comparisons have no implications for demand behavior; as an argument against interpersonal comparisons, the persuasiveness of Samuelson's assertion depends on equating "observable behavior" with "observable demand behavior." A broader interpretation of "observable behavior"—for example, one that regards verbal statements or facial expressions as observable—allows more scope for "meaningful" interpersonal comparisons.

Recently, the pendulum of professional opinion among economists has

---

[11] Cooter and Rappoport [1984] discuss the transformation in economic thought that marked the "ordinalist revolution."

begun to swing back from the positivist view that interpersonal comparisons are meaningless. In an influential paper, Arrow [1971, p. 409] suggests "a reconsideration of the whole concept of equality . . . by reverting to the utilitarian concepts of Bentham, as applied more precisely to the economic realm by F. Y. Edgeworth." Conceding that "the utilitarian approach is not currently fashionable, partly for the very good reason that interpersonally comparable utilities are hard to define," he continues "nevertheless, no simple substitute has yet appeared, and I think it will be useful to pursue this line of study for whatever clarification it will bring." Arrow's concern, however, is with clarifying alternative notions of equality under the assumption that utilities are interpersonally comparable, not with clarifying the meaning of interpersonally comparable utilities.

The rehabilitation of interpersonal comparisons in economics, to the extent that it has taken place, has thus far been formal, not substantive: the validity of interpersonal comparisons has been assumed, not deduced. With the notable exception of Sen [1985, 1987], few economists have discussed the possible meanings of interpersonal comparisons.

Some modern economists who do not dismiss interpersonal comparisons as meaningless regard such comparisons as "normative" rather than "positive." Thus, Fisher [1987, p. 520], discussing two individuals with identical indifference maps, asserts that "we cannot know that the 'true' utility value associated with a given indifference curve is not much higher for one consumer than for another. Indeed, there is no operational meaning to the proposition that it is higher..." Fisher thus appears to endorse the view that interpersonal comparisons are meaningless, but he goes on to argue that the claim that two consumers are equally well off is really a claim that "as a matter of distributive ethics" they "ought" to be treated as equally well off.

It is sometimes claimed that social choice theory requires interpersonal comparisons, but this claim, even if it were true, would not legitimate interpersonal comparisons. Arrow's "impossibility" theorem, which asserts that a dictatorial collective choice rule is the only procedure for aggregating individual preferences consistent with a set of apparently appealing axioms, stands at the center of modern social choice theory. Because one of Arrow's axioms implicitly rules out interpersonal comparisons, Arrow's impossibility or dictatorship theorem is often viewed as demonstrating the limitations of welfare economics without interpersonal comparability. But interpersonal comparability is not the only way of avoiding Arrow's dictatorship conclusion; it can be avoided, for example, by relaxing another axiom such as unrestricted domain or universal scope.[12] Furthermore, even if interpersonal comparisons offered the only possible way of avoiding

---

[12] Blair and Pollak [1983] and Sen [1970, 1982] survey the social choice literature and discuss this and other ways of circumventing the Arrow dilemma.

the Arrow impasse, this would provide no reason for thinking that such comparisons are valid.

We are agnostic about the possibility of making interpersonal welfare comparisons. We find the most convincing argument that such comparisons are possible is the frequency with which they are made. The assertion that individual A is better off than individual B is often understood as a claim about A and B. It is not, for example, necessarily equivalent to the statement that the speaker, on the basis of his or her own preferences, would rather be in A's shoes than in B's. Careful analysis of the conditions under which claims involving interpersonal comparisons are made and command assent or provoke dissent provides insight into their meaning, but economists lack the expertise to analyze the language of welfare comparisons. Despite our inability to provide a satisfactory account of interpersonal comparisons, we are unwilling to dismiss them as a priori impossible or clearly meaningless.[13]

Interpersonal comparisons, if they can be made at all, require a "welfare correspondence" between the indifference maps of two individuals, that is, a correspondence that is appropriate for making interpersonal comparisons. Finding mathematical correspondences between two indifference maps is easy; establishing that a particular correspondence is a welfare correspondence is hard. For example, the simplest mathematical correspondence involving two individuals with identical indifference maps is one that associates with each indifference curve on one map the identical curve on the other. Yet even with identical indifference maps, we cannot conclude that individuals on "the same" indifference curves are equally well-off unless the individuals are equally efficient "pleasure machines." The pleasure machine metaphor, although unappealing, emphasizes the need for a theory of interpersonal comparisons: any coherent attempt to make interpersonal comparisons must rest on some underlying theory, and the theory should not be implicit but explicit.

The lack of an underlying theory of interpersonal comparisons poses a fundamental problem for the mathematical correspondence approach: without a theory, any correspondence selected is arbitrary. Many procedures enable us to establish mathematical correspondences between pairs of indifference maps. For example, given any indifference map, we can draw a 45° ray from the origin and number the indifference curves by their distance from the origin along the 45° ray. Using this procedure to assign numbers to the indifference curves of two individuals, we can define a correspondence between any two indifference maps by associating with each curve on the first map the curve on the second map that has been assigned the same number. The arbitrariness of this correspondence is highlighted by the observation that a different mathematical

---

[13] Other social scientists and philosophers are generally less skeptical than economists about interpersonal comparisons.

correspondence would generally be obtained if the goods were measured in different units (e.g., if the wine consumption of both individuals were measured in gallons rather than in liters) or if the distance from the origin were measured along a different ray.

When individuals have different indifference maps, more elaborate relationships between them can sometimes be found, and these can be used to define a mathematical correspondence between the maps. For example, it is possible that two individuals would have identical maps if one individual's consumption vector were subjected to a simple transformation, for example, divided by a scale factor, δ. In this case, if individual A's indifference map can be represented by the utility function $U(x_1, ..., x_n)$, then individual B's can be represented by the utility function $U(x_1/\delta, ..., x_n/\delta)$. If utility functions were production functions, then we could conclude that individual A with the consumption vector $X^* = (x_1^*, ..., x_n^*)$ would be as well off as individual B with the consumption vector $X^*/\delta = (x_1^*/\delta, ..., x_n^*/\delta)$. But, as the saying goes, "if my grandmother had wheels, she'd be a trolley car."

"Equivalence scales" in empirical demand analysis often rest on the assumption that the indifference maps of different individuals would be identical if consumption of each good were divided by an appropriate scale factor. Thus, if individual A's indifference map can be represented by the utility function $U(x_1, ..., x_n)$, then individual B's can be represented by the utility function $U(x_1/\delta_1, ..., x_n/\delta_n)$, where δ_i is the equivalence scale for good i. The assumption that the indifference maps of households with different demographic profiles are related in this or some other specific way permits the pooling of data from such households in demand estimation. But such equivalence scales are worthless for welfare comparisons: in the absence of a theory of interpersonal comparisons, such equivalence scales lack welfare significance.

Functional form restrictions on preferences such as separability or homogeneity offer another way of defining correspondences between the indifference maps of individuals with different preference orderings. Direct additivity, because of its familiarity and simplicity, provides a clear illustration. Recall from Chapter 2, Section 2, that an individual's preferences are said to exhibit direct additivity if they can be represented by a direct utility function of the form:

$$(35) \qquad\qquad U(X) = \sum u^k(x_k).$$

Recall also that utility functions are merely "representations" of preference orderings, that is, real valued functions that assign higher numbers to preferred baskets of goods. Utility functions are "ordinal" in the sense that if a particular function represents a preference ordering, then every increasing transformation of that function represents the same preference ordering. For example, the preference ordering whose indifference map corresponds to the Cobb–Douglas isoquant map can be represented by the direct utility function

(36) $$U(X) = \prod_k x_k^{\alpha_k}$$

as well as by any increasing transformation of it, for example,

(37) $$V(X) = \sum \alpha_k \log x_k.$$

In Chapter 2 we used the phrase "additive canonical form" to describe any additive representation of the form (35). Because any increasing linear transformation of (35) is also an additive canonical form, such forms are said to be unique up to an increasing linear transformation.

In conjunction with a "normalization rule," the additive canonical form can be used to assign numbers to indifference curves. Because the canonical form is unique up to an increasing linear transformation, we may arbitrarily specify its value at two points. For example, we could let the two points be $X^a = (1, \ldots, 1)$ and $X^b = (2, \ldots, 2)$ and set $U(X^a) = 1$ and $U(X^b) = 2$; with this normalization rule, the additive canonical form is uniquely determined, that is, it defines a unique assignment of numbers to commodity bundles. After assigning numbers to the indifference curves of the two individuals in this way, we can define a correspondence between the two maps by associating with each curve on the first map the curve on the second map with the same number.

The fundamental difficulty with basing interpersonal comparisons on the normalized additive canonical form is the absence of any legitimate reason for doing so. Using the normalized additive canonical form rather than some increasing transformation of it (e.g., its cube or fifth power) to make welfare comparisons is not a harmless simplification and, hence, mathematical convenience provides no justification for its use. Instead, its justification requires an argument that the additive canonical form is appropriate for making interpersonal comparisons.

This objection to basing interpersonal comparisons on the additive canonical form also precludes basing such comparisons on any other mathematically specified canonical form. This is immediately clear in the case of indirect additivity and only slightly less clear in the case of the Gorman polar form, Muellbauer's PIGLOG form, or any other canonical form specified in terms of the way total expenditure enters the demand system. Basing welfare comparisons on any of these canonical forms is vulnerable to the same objection as using the additive canonical form. There is no rationale for preferring welfare comparisons based on these canonical forms to those based on any other representation of preferences. Welfare comparisons based on canonical forms are arbitrary.

Arrow [1951] dismissed the possibility of using individual von Neumann–Morgenstern utility functions to avoid his impossibility result, arguing that it is inappropriate to allow individuals' tastes for gambling to affect social outcomes. Regardless of whether one agrees with Arrow's conclusion—and an argument can be made that individual risk preferences should play a role in aggregating individual preferences into social preferences or social choices—Arrow is certainly correct that the appropriateness of using von Neumann–Morgenstern utility functions as a basis for preference aggregation or welfare comparisons requires arguing that they have welfare significance.[14] Indeed, Arrow's point has broader applicability: the appropriateness of using any particular form of the utility function (i.e., any representation of the preference ordering) for welfare comparisons depends on its welfare significance.

Even if we accept the notion that attitudes toward risk are relevant to welfare comparisons, however, von Neumann–Morgenstern utility functions do not contain enough information to permit such comparisons. Using vonNeumann–Morgenstern utility functions to make interpersonal comparisons involves some but not all of the same problems as using the additive canonical form. Because von Neumann–Morgenstern utility functions are unique up to a linear transformation, a normalization rule is required to obtain a particular numbering of an individual's indifference curves. To justify interpersonal comparisons based on the numerical value of the von Neumann–Morgenstern utility function corresponding to a particular normalization rule requires a normalization rule appropriate for making interpersonal comparisons, and identifying such a normalization rule requires a theory of interpersonal comparisons.

2.3. Households and Families

Theoretical statements of the situation and welfare comparison problems are individualistic because preferences are individualistic: situation comparisons require a single preference ordering, and welfare comparisons require two preference orderings and a correspondence between them. Practical statements of both the situation and welfare comparison problems, however, usually are framed in terms of households or families.[15] Thus, to make situation comparisons or welfare comparisons involving households or families, we must either reinterpret preferences in terms of households or families or reformulate welfare and situation comparisons so that they do not require preferences.

If we accept the notion of household or family preferences—a move Nerlove [1974] calls the "Samuelson finesse" in recognition of Samuelson's [1956] insightful discussion in which he argued that aggregating the preferences of the individuals in a family and aggregating the preferences of the individuals in a society are essentially identical problems—then comparisons based on family preferences present no new difficulties. There

---

[14] Harsanyi [1953, 1955] advances two distinct arguments for using von Neumann–Morgenstern utility functions in making welfare evaluations.
[15] Distinguishing between households and families is sometimes important, but it is not crucial here.

is, however, much to be said for rejecting the notion of household or family preferences and, for that matter, any notion of preferences other than individual preferences.[16] Indeed, those who think that making interpersonal welfare comparisons is nonsense are also likely to think that making interfamily welfare comparisons is—to borrow a phrase from Bentham—"nonsense on stilts."

Economists usually lack direct observations on individual consumption patterns, but neither individual consumption nor individual preferences can be recovered from observations on household purchases except under highly restrictive assumptions. Although all empirical analysis rests on assumptions as well as on data, estimating intrahousehold allocation and individual preferences from data on household purchases places an especially heavy burden on assumptions. Lazear and Michael [1988] attempt to estimate individual consumption patterns from data on the purchase patterns of households with different demographic profiles and Gronau [1988] offers a theoretical interpretation of equivalence scales in terms of individual preferences. A serious limitation of these analyses is the absence of a satisfactory treatment of household public goods. Blackorby and Donaldson [1989] derive strong conclusions about interpersonal comparisons under the assumption that the household social welfare function maximizes the utility of the worst-off individual, but offer no justification for their strong assumption about the form of the household social welfare function.

The contrast between theoretical formulations involving individuals and practical ones involving households or families is pervasive in economics. Although individual preferences are central in the theory of consumer behavior and welfare economics, households and families are the principal focus of empirical demand analysis and public policy discussions. For an important range of social and economic issues (e.g., marriage, divorce, child support, intrafamily allocation) the focus on households or families rather than individuals is clearly inappropriate. For other issues (e.g., demand analysis using household level data) the use of the household as the unit of analysis is so familiar in economics that it seems "natural" and, hence, the transition from individual preferences to household or family preferences is usually made without discussion or acknowledgment. Recent theortical work on bargaining models of marriage (e.g., Manser and Brown [1980] and McElroy and Horney [1981]) challenges the notion of household or family preferences (see Pollak [1985]). Recent empirical work on the labor supply of husbands and wives (Lundberg [1988]) challenges the appropriateness of models postulating "family preferences."

---

[16] Thus, many theorists object to Arrow's formulation of the social choice problem as one of aggregating individual preferences into social preferences; but even when the social choice problem is reformulated in terms of social choices rather than social preferences, the Arrow paradox remains.

Abandoning the notion of household or family preferences exposes two distinct issues: the treatment of conflicting preferences of adults within the household, and the treatment of children's interests, needs, wants, and desires. When these issues are taken seriously, as they deserve to be, the question, "How much would a household or family consisting of a man, a woman, and three children need to be as well off as a household or family consisting of a man, a woman, and two children and $20,000?" exemplifies the double-barreled complexity of welfare comparisons involving households or families.

2.4. Conclusion

Situation comparisons are firmly grounded in economic theory. The theory of the cost-of-living index provides a theoretical framework for comparing alternative price situations on the basis of a single preference ordering. The generalized cost-of-living index extends this framework to permit comparisons of situations differing not only in prices but in other dimensions as well (e.g., nonmarket goods, demographic variables).

Although the generalized cost-of-living index provides a theoretical foundation for making situation comparisons, its construction requires information about unconditional preferences that is difficult to obtain. Using revealed preference reasoning, we can recover unconditional preferences from unconditional choices: for example, if an individual chooses one goods–demographic situation when another goods–demographic situation is available, then we infer that the individual prefers the first situation to the second. We cannot, however, recover unconditional preferences from conditional choices: for example, if the individual takes the demographic profile as fixed and chooses among alternative bundles of goods, we cannot recover the individual's unconditional preferences from these data.

Welfare comparisons require more information than situation comparisons. To compare the welfare of two individuals, we must know their unconditional preferences and we must know the welfare correspondence between them. Establishing the welfare correspondence between two individuals' indifference maps—that is, determining which indifference curve on one individual's map yields the same welfare level as a particular curve on the other's map—presupposes a theory of interpersonal comparisons and economics lacks such a theory.[17]

"Exact aggregation" is a key component of recent attempts by Jorgenson and others to make comparisons involving households with different

---

[17]When individual or household welfare depends on demographic variables as well as the consumption of goods, then the measurement of poverty (see Sen [1976]) and of inequality (see Sen [1973], Maasoumi [1986]) as well as the construction of the social cost-of-living index (see Pollak [1981], Jorgenson and Slesnick [1983]) requires that we distinguish between conditional and unconditional preferences.

demographic profiles. A lineal descendant of the Gorman polar form and PIGLOG, exact aggregation implies restrictions on individuals' conditional demand functions that simplify the aggregate demand functions. In Jorgenson's [1990, p. 5] words: "Under exact aggregation the aggregate demand system depends on summary statistics of the joint distribution of attributes and total expenditure among individuals."[18]

We are now positioned to assess the usefulness of exact aggregation as a basis for making situation comparisons and welfare comparisons. Situation comparisons require unconditional preferences, yet exact aggregation, with its focus on conditional choices, is not a plausible procedure for recovering unconditional preferences. Exact aggregation restricts conditional preferences to a particular parametric class, but even with this restriction, there is no reason to think that unconditional preferences involving such variables as family size and race can be recovered from differences in the consumption patterns of households with different demographic profiles. Welfare comparisons require not only unconditional preferences but also a welfare correspondence between indifference maps. The meaning of welfare comparisons is ambiguous, but even when the conditional preferences of these households are restricted to a particular parametric class there is no reason to think that an appropriate welfare correspondence can be deduced from differences in the consumption patterns of households with different demographic profiles.

The generalized cost-of-living index provides an elegant answer to the situation comparison question, not to the welfare comparison question. It is tempting to abandon the intractable welfare comparison problem and substitute for it the less intractable situation comparison problem. This temptation should be resisted: the two problems are not equivalent. Welfare comparisons require making interpersonal comparisons, not avoiding them.

APPENDIX A: PROOFS OF THEOREMS 1 and 2

Proof of Theorem 1: It suffices to show that the implied demand functions correspond to the utility function $U(X)$. To show this, we observe that the original demand functions are solutions to the first order conditions:

(A1)
$$u^{i'}(x_i) = -\lambda p_i$$

$$\sum p_k x_k = \mu.$$

---

[18]Exact aggregation was introduced by Lau [1977], and used as a basis for welfare comparisons by Jorgenson, Lau, and Stoker [1980] and Jorgenson and Slesnick [1983]. Jorgenson [1990] summarizes and provides references to this line of work.

The modified demand functions are defined in terms of the original demand functions by

(A2) $$h^i(P, \mu) = s_i \bar{h}^i(P, \mu/s_o)$$

where s_o is implicitly defined by

(A3) $$\sum p_k s_k \bar{h}^k(P, \mu/s_0) = \mu.$$

We must show that these modified demand functions satisfy the first order conditions corresponding to the utility function $U(X)$, namely:

(A4)
$$u^{i'}(x_i s_i) = \lambda p_i$$
$$\sum p_k x_k = \mu.$$

That is, we must show that the modified demand functions satisfy

(A5)
$$u^{i'}[h^i(P, \mu/s_i)] = -\lambda p_i$$
$$\sum p_k h^k(P, \mu) = \mu,$$

or, equivalently

(A6)
$$u^{i'}[\bar{h}^i(P, \mu/s_0)] = -\lambda p_i$$
$$\sum p_k s_k \bar{h}^k(P, \mu/s_o) = \mu.$$

The original demand functions $\{\bar{h}^i(P, \mu/s_o)\}$ maximize the utility function $\bar{U}(X)$ subject to the budget constraint $\sum p_k x_k = \mu/s_o$, so they satisfy the first order conditions

(A7)
$$u^{i'}[\bar{h}^i(P, \mu/s_o)] = -\lambda p_i$$
$$\sum p_k \bar{h}^k(P, \mu/s_o) = \mu$$

where s_o is defined implicitly by

(A8) $$\sum p_k s_k \bar{h}^k(P, \mu/s_o) = \mu.$$

But this means that the condition (A6) is satisfied.   QED

Proof of Theorem 2: The proof depends on establishing that the conditions for the existence of theoretically plausible demographic demand functions are identical to those for the existence of theoretically plausible long-run demand functions in a linear habit formation model. This permits us to make immediate use of a theorem from Pollak [1976b, p. 284] characterizing this class of functions.

If the original demand system is theoretically plausible and linear in expenditure, then the modified Prais–Houthakker procedure yields demand functions of the form given in (17). We write (17) as

(A9) $$h^i(P, \mu) = B^i(P) - \Gamma^i(P)\sum p_k B^k(P) + \Gamma^i(P)\mu$$

where $B^i(P)$ and $\Gamma^i(P)$ are defined by

(A10) $$B^i(P) = \frac{f_i(P)}{1 - \beta_i}, \qquad \Gamma^i(P) = \frac{\gamma^i(P)/(1 - \beta_i)}{\sum p_k \gamma^k(P)/(1 - \beta_k)}$$

and β_i by $\beta_i = 1 - 1/s_i$; hence $s_i = 1/(1 - \beta_i)$. Except for the absence of the b*'s in the expressions for the B's, (A9) is precisely the form of the long-run demand functions given in Pollak [1976b, Eq. (3.7)]. Since variations in the b*'s play no role in the proof of the theorem (Pollak [1976b, p. 284]), it follows that our demographic demand functions are theoretically plausible if and only if the original demand system satisfies the conditions established in Pollak [1976b], and these conditions are equivalent to the requirement that the original demand system correspond to an additive direct utility function.   QED

4

Dynamic Structure

Economists have traditionally been suspicious of changing tastes, and intellectual tastes change slowly.[1] Nevertheless, it is time to reconsider the conventional wisdom that tastes are not the business of economists.

Those who favor incorporating taste formation and taste change into economic analysis fall into two groups whose intersection is virtually empty. One group is primarily interested in the welfare implications of changing tastes, the other in the analysis of household behavior. Galbraith, who emphasizes the ability of producers to manipulate consumers through advertising, provides an articulate statement of the welfare view. Characterizing his own work, he writes, "The surrender of the sovereignty of the individual to the producer or producing organization is the theme, explicit or implicit, of two books, *The Affluent Society*... and *The New Industrial State*" [1970, p. 471]. Veblen's notion of "conspicuous consumption" suggests a model in which preferences for goods depend directly on prices because people judge quality by price or because a higher price enhances "snob appeal." Pollak [1977] discusses price-dependent preferences and provides references to the literature. "Radical" economists of both Marxist and non-Marxist persuasions also reject the notion that taste formation and change are outside the province of economics or else deny that economics can be separated from the other social sciences (see, for example, Gintis [1974]).

The recent impetus to incorporate taste formation and change into economic analysis, however, has come primarily from those interested in household behavior rather than welfare, and the principal focus of this work has been empirical demand analysis. In Chapter 1 we pointed out that empirical demand analysis must either assume that all demand system parameters remain constant over time or specify how they change. We showed that the LES could be given a dynamic structure by allowing the b's to depend linearly on consumption in the previous period and interpreted that dynamic structure as habit formation. In Section 1 of this chapter we extend that analysis to other specifications of habit formation and to other demand systems, examining the implied short-run and

---

[1] This chapter is based on Pollak [1970, 1976a, 1976b, 1978, 1990].

long-run behavior. In Section 2 we discuss interdependent preferences, an alternative dynamic specification in which preferences and demand depend on the consumption patterns of others. Finally, in Section 3 we discuss briefly the implications of these dynamic specifications for the evaluation of welfare.

1. HABIT FORMATION

The LES

$$(1) \qquad h^i(P, \mu) = b_i - \frac{a_i}{p_i} \sum p_k b_k + \frac{a_i}{p_i} \mu$$

is generated by the direct utility function

$$(2) \quad U(X) = \sum a_k \log(x_k - b_k), \qquad a_i > 0, \qquad (x_i - b_i) > 0, \qquad \sum a_k = 1.$$

In Chapter 1 we introduced a simple model of habit formation in which the b's depend linearly on consumption in the previous period

$$(3) \qquad b_{it} = b_i^* + \beta_i x_{it-1}.$$

Under this specification the utility function becomes

$$(4) \qquad U(X_t; X_{t-1}) = \sum a_k \log(x_{kt} - b_k^* - \beta_k x_{kt-1})$$

and the corresponding short-run demand functions

$$(5) \quad h^i(P_t, \mu_t; X_{t-1}) = b_i^* - \frac{a_i}{p_{it}} \sum p_{kt} b_k^* + \frac{a_i}{p_{it}} \mu_t + \beta_i x_{it-1} - \frac{a_i}{p_{it}} \sum p_{kt} \beta_k x_{kt-1}.$$

Because the b's are linear in past consumption and because current consumption depends linearly on the b's, these short-run demand functions imply that present consumption of each good is a linear function of past consumption of all goods. Because the β's are positive, an increase in past consumption of a good implies an increase in current consumption of that good

$$(6) \qquad \frac{\partial h^{it}(P, \mu; X_{t-1})}{\partial x_{it-1}} = \beta_i - a_i \beta_i > 0$$

and a decrease in current consumption of every other good

$$(7) \qquad \frac{\partial h^{jt}(P, \mu; X_{t-1})}{\partial x_{it-1}} = a_j p_i \beta_i / p_j < 0, \qquad i \neq j.$$

When the b's in the LES are positive, we can interpret them as a necessary collection of goods, but the utility function (2) and the demand functions (1) are defined for negative as well as positive b's. Even if b_i^* and b_{it} are negative, the short-run demand functions retain all of the general properties described above. The only casualty is the interpretation of b_i^* and b_{it} as components of a necessary basket. The habit hypothesis, however, does not require this interpretation. The two essential features of habit formation are: first, that past consumption influences current preferences and, hence, current demand; second, that a higher level of past consumption of a good implies, ceteris paribus, a higher level of current consumption of that good. It is easily verified that these conditions are satisfied by (4) regardless of the signs of b_{it} and b_i^*.

The habit formation specification (3) implies that consumption in the previous period influences current preferences and demand but that consumption in the more distant past does not. An example of a specification that allows consumption in the more distant past to matter is one in which the necessary quantity of each good depends on a geometrically weighted average of all past consumption of that good. The analogue of (3) is

$$(8) \qquad\qquad b_{it} = b_i^* + \beta_i z_{it-1}$$

where

$$(9) \qquad\qquad z_{it-1} = (1 - \delta_i) \sum_{\tau=0}^{\infty} \delta_i^\tau x_{it-1-\tau}.$$

The δ's are "memory" coefficients. The one-period lag specification, (3), corresponds to the special case in which all $\delta_i = 0$. Substituting (8) into (1), we obtain a short-run demand system that depends on all past levels of consumption, not just on consumption in the previous period. This demand system is of the same form as (4) except that x_{it-1} is replaced by z_{it-1}. Because z_{it-1} depends linearly on past consumption of the ith good, it is easy to show that, ceteris paribus, a higher level of past consumption of a good implies a higher level of current consumption of that good.

An alternative habit hypothesis is that the b's depend linearly on the previous peak consumption of a good:

$$(10) \qquad\qquad b_{it} = b_i^* + \beta_i \max_{\tau<t} x_{i\tau}.$$

This implies an irreversible ratchet similar to that suggested by Duesenberry [1949] and Modigliani [1949] in the consumption function context. The long-run irreversibility of the ratchet is implausible because it implies that the effect of previous peak consumption on current demand is independent of how long ago the peak occurred. There are several ways of avoiding this irreversibility while maintaining the appealing features of the ratchet hypothesis. One is to assume that tastes are influenced by the highest consumption level attained in the previous t* periods:

$$(11) \qquad\qquad b_{it} = b_i^* + \beta_i \max_{\tau-t^*<\tau<t} x_{i\tau}.$$

Another is to assume that the influence of the previous peak diminishes with the passage of time, an assumption that avoids the need to specify a fixed horizon beyond which everything is forgotten and before which everything is recalled perfectly. The implied habit function is of the form

$$(12) \qquad b_{it} = b_i^* + \beta_i(\delta_i)^{t-\tau} \max_{\tau < t} x_{i\tau}$$

where δ_i is again the memory coefficient for the ith good. An alternative formalization assumes the habit function depends on the previous peak of perceived consumption rather than actual consumption. If the memory of past consumption decays geometrically, then consumption in the distant past is "discounted" more heavily than consumption in the recent past and the ratchet is reversible:

$$(13) \qquad b_{it} = b_i^* + \beta_i \max_{\tau < t} (\delta_i)^{t-\tau} x_{i\tau}.$$

We have thus far ignored the simplest dynamic specification, the linear time trend,

$$(14) \qquad b_{it} = b_i^* + \beta_i t.$$

Time trend specifications have three serious drawbacks. First, time trends give no insight into the social or economic forces generating taste change. In contrast, the lagged consumption specification may be interpreted as habit formation or interdependent preferences. Second, time marches on and the time trend specification implies inexorable taste change even if prices and expenditure remain constant for many periods. We find this implausible. Third, in conjunction with the LES, the generalized CES, and some other widely used functional forms, time trends applied to the necessary quantities imply an eventual violation of regularity conditions: the difficulty is that if prices and expenditure remain constant, then the value of the necessary basket will eventually exceed expenditure. For these reasons, we prefer dynamic specifications based on past consumption.

Our assumptions about the form of the habit function can be generalized in three directions. First, the habit function can be nonlinear in past consumption

$$(15) \qquad b_{it} = H^i(x_{it-1}),$$

where $H^i(x_{it-1})$ denotes a continuous, increasing function. Second, the habit function for each good can depend not only on past consumption of that good but also on past consumption of other goods:

$$(16) \qquad b_{it} = H^i(x_{it-1}, \ldots, x_{nt-1}).$$

Third, as we have already seen, the habit function can depend not only on consumption in the previous period but also on consumption in the more distant past:

(17) $$b_{it} = H^i(X_{t-1}, X_{t-2}, \ldots, X_{t-\tau}, \ldots).$$

Because these generalizations introduce no new conceptual issues, we have chosen to emphasize the most transparent specification—the one in which the habit function for each good depends linearly on consumption of that good in the previous period.

The specifications we have considered thus far assume that some but not all of the LES parameters depend on past consumption. Because the demand functions are linear in the b's, a specification in which the b's depend linearly on past consumption yields dynamic demand systems that are linear in past consumption. It might appear that this is also true of a specification in which the a's depend linearly on past consumption, but it is not. Unlike the b's, which can be specified independently, the a's must sum to 1: $\sum a_k = 1$. Hence, a change in past consumption causing an increase in one of the a's must cause offsetting decreases in the others.

The issues that arise in applying alternative dynamic specifications are similar to those that arise in applying demographic specifications to arbitrary demand systems. This is not surprising because in both instances we seek to incorporate additional explanatory variables. There are, however, two significant differences. First, dynamic specifications typically involve one lagged consumption variable for each good, while demographic specifications must be capable of accommodating both a single explanatory variable (e.g., family size) and a large number of such variables (e.g., race, ages, and sexes of family members, place of birth, etc.). Second, dynamic specifications often postulate a close association between the lagged and current consumption of each good; with demographic specifications there is often little reason to postulate a close association between the consumption of a particular good and a particular demographic variable, even when the number of demographic variables happens to be equal to the number of goods.

A general procedure for obtaining a dynamic specification of an arbitrary demand system is to allow some or all of its parameters to depend on past consumption. Theory provides little guidance in specifying the parameters that depend on past consumption, the variables that represent past consumption, or the functional form that the habit function assumes. Even if we confine ourselves to dynamic specifications that depend only on consumption in the previous period, each demand system parameter may depend on the lagged consumption of all goods. Two kinds of restrictions are implied by the theory. First, and most important, are those that follow from the fact that the demand system parameters themselves are often not independent (e.g., in the LES, the a's must sum to unity). Second, habit formation postulates a positive relationship between past and current consumption of each good.

Dynamic translating and dynamic scaling single out particular subsets of demand system parameters and assume that these and only these parameters depend on past consumption. Both translating and scaling are general procedures in the sense that they can be used in conjunction with any original demand system, $\{x_i = \bar{h}^i(P,\mu)\}$. As in Chapter 3, we assume that these original demand systems are theoretically plausible, and denote the corresponding direct and indirect utility functions by $\bar{U}(X)$ and $\bar{\psi}(P,\mu)$.

There are two versions of dynamic translating. If the original demand system contains constant terms, then we assume that the constants depend on past consumption. In principle, the constants may depend on any variables representing past consumption, but we assume that they depend only on the previous period's consumption. Log linear dynamic translating, one of the specifications we estimate in Chapter 7, is given by

$$(18) \qquad D^i(x_{it-1}) = d_i x_{it-1}^{\gamma_i}$$

and adds n parameters to the original demand system.

When the original demand system does not contain constant terms (e.g., in the Cobb–Douglas case), dynamic translating introduces them by replacing the original system by

$$(19) \qquad h^i(P,\mu) = b_i + \bar{h}^i(P, \mu - \textstyle\sum p_k b_k).$$

These newly introduced constants are then assumed to depend on past consumption. In this case, the dynamic demand system does not contain constant terms independent of past consumption unless all of the γ's are 0. A more general formulation avoids this difficulty by retaining the constant term:

$$(20) \qquad D^i(x_{it-1}) = b_i^* + d_i x_{it-1}^{\gamma_i}.$$

If the original demand system is theoretically plausible, then the modified system is also, at least for d's close to 0. The modified system satisfies the first order conditions corresponding to the indirect utility function $\psi(P,\mu) = \bar{\psi}(P, \mu - \sum p_k d_k)$. With linear translating, provided $\partial D^i/\partial x_{it-1} > 0$, an increase in past consumption of a good implies an increase in current consumption of that good provided that the good's marginal budget share is less than 1:

$$(21) \qquad \frac{\partial h^i}{\partial x_{it-1}} = \frac{\partial D^i}{\partial x_{it-1}} \left[1 - p_i \frac{\partial \bar{h}^i}{\partial \mu} \right].$$

Again provided $\partial D^i/\partial x_{it-1} > 0$, the increase in past consumption of a good implies a decrease in current consumption of every noninferior good:

$$(22) \qquad \frac{\partial h^j}{\partial x_{it-1}} = - \frac{\partial \bar{h}^j}{\partial \mu} p_i \frac{\partial D^i}{\partial x_{it-1}}.$$

Dynamic translating was originally proposed by Stone [1954, p. 552] in the context of the LES; he implemented his own proposal in Stone [1966a, 1966b]. Houthakker and Taylor [1966, 2nd ed. 1970] proposed and estimated a model in which past consumption influences consumption patterns through a "state" variable which they interpret as a "psychological stock" of habits. In their second edition they showed how such a dynamic demand system can be obtained from utility maximization. Theoretical models of habit formation are investigated in Gorman [1967], Peston [1967], Pollak [1970, 1976b], von Weizsäcker [1971], Gaertner [1974], Lluch [1974], McCarthy [1974], Phlips [1974, rev. ed. 1983], El-Safty [1976a, 1976b], Hammond [1976], Klijn [1977], Spinnewyn [1981], Boyer [1983], Pashardes [1986], and Becker and Murphy [1988]. Empirical investigations based on various specifications of habit formation include Pollak and Wales [1969, 1987], Wales [1971], Phlips [1972], Brown and Heien [1972], Taylor and Weiserbs [1972], Boyce [1975], Manser [1976], Anderson and Blundell [1983], and Darrough, Pollak, and Wales [1983]. Blundell [1988] provides a recent survey.

Dynamic scaling, an alternative dynamic specification, replaces the original demand system by

$$(23) \qquad h^i(P, \mu) = m_i \bar{h}^i(p_1 m_1, \dots, p_n m_n, \mu),$$

where the m's are scaling parameters that depend on the previous period's consumption: $m_i = M^i(x_{it-1})$. A more general formulation would allow all m_i to depend on consumption in all previous periods. If the original demand system is theoretically plausible, then the modified system is also, at least for m's close to 1. The modified system satisfies the first order conditions corresponding to the indirect utility function $\psi(P, \mu) = \bar{\psi}(p_1 m_1, \dots, p_n m_n, \mu)$ and the direct utility function $U(X) = \bar{U}(x_1/m_1, \dots, x_n/m_n)$. Loosely speaking, we can interpret x_i/m_i as a measure of x_i in "efficiency units" rather than in physical units. The effect of an increase in past consumption of good i on current consumption of good i is given by

$$(24) \qquad E^i_{it-1} = \hat{M}^i_{it-1} + E^i_i \hat{M}^i_{it-1},$$

where E^i_{it-1} is the elasticity of demand for good i with respect to lagged consumption of good i and $\hat{M}^i_{it-1}$ is the elasticity of m_i with respect to x_{it-1}. The effect of an increase in past consumption of good i on current consumption of good j is given by

$$(25) \qquad E^j_{it-1} = E^j_i \hat{M}^i_{it-1}.$$

In certain special cases (e.g., the LES) dynamic translating and dynamic scaling coincide.

Log linear dynamic scaling, one of the specifications we estimate in Chapter 7, is given by

$$(26) \qquad M(x_{it-1}) = x^{\gamma_i}_{it-1}.$$

This specification, which adds n parameters to the original demand system, guarantees that the implied value of m_i will be positive, as theory requires.

We now consider the long-run behavior corresponding to a short-run demand system. Given the consumption vector of period 0, and given prices and expenditure of period 1, the short-run demand functions yield a consumption vector for period 1. Suppose prices and expenditure remain the same in period 1 as in period 0. A "steady-state" or "long-run equilibrium" consumption vector is one that, if it prevailed in period 0, would also prevail in period 1. If prices and expenditure remain constant over time, the consumption vector in each subsequent period would also equal the consumption vector of period 0.

Although estimation must be based on the short-run demand system, we are often interested in long-run as well as short-run responses to changes in prices or expenditure. The long-run equilibrium consumption vector can be found by solving the short-run demand functions (5) under the assumption that $x_{it} = x_{it-1} = x_i$ for all i, but an alternative solution procedure is sometimes simpler. As usual, the LES provides an example. The first order maximization conditions corresponding to (4) are

$$(27) \qquad \frac{a_i}{(x_{it} - b_{it})} = \lambda p_i,$$

$$\sum p_k x_{kt} = \mu.$$

In the short run the utility maximizing x's must satisfy (27), where the b's are determined by past consumption. In the long-run equilibrium, however, the b's are given by $b_i = b_i^* + \beta_i x_i$ where x_i is the long-run equilibrium value of x_{it}. Thus, in the long-run equilibrium the x's must satisfy

$$(28) \qquad \frac{a_i}{(x_i - b_i^* - \beta_i x_i)} = \nu p_i,$$

$$\sum p_k x_k = \mu,$$

where ν represents the long-run value of the Lagrangian multiplier. In the short run, the value of the Lagrangian multiplier depends on the values of the b's and, hence, on past consumption as well as on prices and expenditure. In the long run, however, the values of x_{it} and x_{it-1} must be equal, and the long-run value of the Lagrangian multiplier depends only on prices and expenditure. Solving the "long-run first order conditions," (28), for x_i yields

$$(29) \qquad x_i = \frac{b_i^*}{1 - \beta_i} + \frac{a_i}{1 - \beta_i} \frac{1}{\nu} \frac{1}{p_i}.$$

Multiplying (29) by p_i, summing over all goods, solving for $(1/\nu)$, and substituting into (29), we obtain the "long-run" or "equilibrium" demand functions

(30)
$$h^i(P, \mu) = B_i - \frac{A_i}{p_i} \sum p_k B_k + \frac{A_i}{p_i} \mu,$$

where

(31)
$$A_i = \frac{a_i/(1 - \beta_i)}{\sum a_k/(1 - \beta_k)}, \qquad B_i = \frac{b_i^*}{1 - \beta_i}.$$

These long-run demand functions show the steady-state consumption patterns consistent with the short-run demand functions (5).

In general, there is no guarantee that a long-run demand system defined as the steady-state or equilibrium values corresponding to the short-run demand system will satisfy the symmetry or negative semidefiniteness conditions, or, equivalently, that they can be "rationalized" by a "long-run utility function." When the short-run demand system is an LES, however, it is obvious that the long-run demand functions (30) are also an LES and can be rationalized by the long-run utility function

(32) $U(X) = \sum A_k \log(x_k - B_k), \qquad A_i > 0, \qquad (x_i - B_i) > 0, \qquad \sum A_k = 1$

where A_i and B_i are defined by (31). Clearly $A_i > 0$ and $\sum A_k = 1$. If $\bar{x}_i$ is an admissible long-run equilibrium, then $[\bar{x}_i - (b_i^* + \beta_i \bar{x}_i)] > 0$, so $(x_i - B_i) > 0$. Since the long-run demand functions (30) can be derived from a well-behaved utility function, they must satisfy the symmetry and negative semidefiniteness conditions.

These long-run LES demand functions were not derived by maximizing a long-run utility function and, in particular, they are not the demand functions implied by maximizing the utility function obtained by replacing b_{it} by $b_i^* + \beta_i x_{it-1}$ in (2):

(33)
$$V(X) = \sum a_k \log [x_k - (b_k^* + \beta_k x_k)].$$

The procedure used here for the LES can be applied directly to the generalized CES, and, with slight modification, to the additive exponential utility function.[2] These forms, together with the LES, exhaust the class of demand systems generated by additive direct utility functions that imply linear Engel curves.

In most cases the long-run demand functions generated as steady-state solutions of a habit formation model cannot be rationalized by a long-run utility function. Pollak [1976b] shows that if the short-run demand functions are locally linear in expenditure, then only the subclass corresponding to additive direct utility functions generates long-run demand functions that can be rationalized by a long-run utility function. El-Safty [1976a, 1976b] shows that separability is the crucial requirement.

---

[2] Pollak [1970] works out the details.

Establishing local dynamic stability of the long-run demand system requires matrix algebra. It is straightforward to write the short-run demand system in matrix form as

$$(34) \qquad\qquad X_t = b_t - \gamma_t P'_t b_t + \gamma_t \mu_t$$

where b_t is given by

$$(35) \qquad\qquad b_t = b^* + \hat{\beta} X_{t-1}$$

and γ^t is a column vector whose elements are $\gamma^{it}(P_t)$; in the case of the LES these are given by

$$(36) \qquad\qquad \gamma^{it}(P_t) = a_i/p_{it}.$$

If X_0 (that is, the consumption vector in period 0) is given, then (34) determines X_1 as a function of P_1, μ_1, and X_0. In the same way, X_2 is determined by (34) as a function of P_2, μ_2, and X_1, or, more conveniently, as a function of $X_0, P_1, \mu_1, P_2, \mu_2$. Thus, for any initial consumption vector X_0 and any price–expenditure sequence $\{(P_1, \mu_1), (P_2, \mu_2),...\}$, (34) determines the corresponding consumption sequence $\{X_1, X_2,...\}$.

We have already identified the equilibrium consumption vector X^* corresponding to the price–expenditure situation (P^*, μ^*). Clearly, if $X_0 = X^*$ and $\{(P_1, \mu_1), (P_2, \mu_2),...\} = \{(P^*, \mu^*), (P^*, \mu^*),...\}$, then $\{X_1, X_2,...\} = \{X^*, X^*,...\}$. To establish local dynamic stability, it is necessary to show that if X_0 is sufficiently close to X^*, then the consumption sequence corresponding to $\{(P^*, \mu^*), (P^*, \mu^*),...\}$ converges to X^*.

With prices and expenditure assumed constant over time, we drop the time subscripts on P_t, μ_t, and γ_t. Substituting (35) into (34) yields

$$(37) \qquad\qquad X_t = MX_{t-1} + d,$$

where

$$(38) \qquad\qquad M = (I - \gamma P')\hat{\beta}$$

and

$$(39) \qquad\qquad d = (I - \gamma P')b^* + \gamma\mu.$$

It is easily verified that X_t is given by

$$(40) \qquad\qquad X_t = M^t X_0 + \left[\sum_{\tau=0}^{t-1} M^\tau\right] d.$$

The stability of this system of difference equations (37) rests on the following theorem.

Theorem: Let M be the matrix defined by (38) where γ and P are $n \times 1$ vectors with positive elements such that $P'\gamma = 1$, and $\hat{\beta}$ is the diagonal matrix diag $(\beta_1,...,\beta_n)$ where $0 \leqslant \beta_i < 1$ for all i; then the characteristic roots of M are all less than 1 in modulus. The proof is given in Pollak [1970, Appendix A].

If the characteristic roots of M are less than 1 in modulus, it is well-known that

(41a) $$\lim_{t \to \infty} M^t = 0$$

and

(41b) $$\lim_{t \to \infty} \sum_{\tau=0}^{t-1} M^\tau = (I - M)^{-1}$$

so the system of difference equations (37) converges to

(42) $$X = (I - M)^{-1} d.^3$$

A rigorous discussion of dynamic stability is inevitably complicated by regularity and nonnegativity conditions that must be satisfied in every period. The local argument is easy: corresponding to every admissible price–expenditure situation, there exists a neighborhood of the long-run equilibrium such that, for initial values of X in this neighborhood, the sequence of consumption vectors $\{X_1, X_2, \ldots\}$ satisfies the nonnegativity and regularity conditions in each period.

This local stability argument can be applied directly to the generalized CES demand system and, with slight modification, to the demand system corresponding to the exponential direct utility function. The analysis in these cases is relatively straightforward because these demand systems, like the LES, are linear in lagged consumption.

Our stability analysis assumes that habit formation depends only on consumption in the previous period. McCarthy [1974] extends the local stability proof to the case in which habit formation depends on a geometrically weighted average of all past consumption.

2. INTERDEPENDENT PREFERENCES

Interdependent preferences—preferences that depend on other people's consumption—have a long history on the margin of economic analysis. The post-Veblen literature on interdependent preferences begins with Duesenberry's well-known book on the consumption function [1949]. Leibenstein [1950] and Prais and Houthakker [1955, Chapter 2] discuss interdependent preferences, but the subject appears to have been dormant from the mid-1950s until the 1970s when it was revived by Krelle [1973], Gaertner [1974], Pollak [1976a], and Hayakawa and Venieris [1977]. Easterlin's "relative income" model of fertility contains elements of both habit formation and interdependent preferences; see Easterlin [1973, 1976],

---

[3] For a discussion of stability see Luenberger [1979, pp. 154–157].

Easterlin, Pollak, and Wachter [1980], and Leibenstein [1975, 1976]. Case [1991] proposes a specification in which demand has a spatial component and points out that her spatial specification can be interpreted in terms of interdependent preferences.

Interdependent preferences can be incorporated into demand analysis using models similar to the habit formation models described in Section 1. We begin by specifying a short-run model of interdependent preferences, "linear interdependence." We introduce linear interdependence in the context of the LES by postulating that the necessary quantities depend linearly on other people's *past* consumption. If the necessary quantities were determined by other people's current consumption, then analyzing the model would require determining everyone's consumption simultaneously. By assuming that interdependent preferences operate through past consumption, we retain the central insight while avoiding peripheral mathematical complications. We consider a number of alternative specifications of interdependent preferences, and extend the results from the LES to more general demand systems. We then aggregate over individuals to obtain the per capita demand functions and investigate the long-run or steady-state equilibrium implied by these per capita demand functions. We emphasize the per capita rather than the individual demand functions because the outcome of the mutual interaction of consumption and tastes that characterizes interdependent preferences is manifested only in the long-run per capita demand functions. Although in models of interdependent preferences one might expect the expenditure distribution to be a significant determinant of per capita consumption patterns, our analysis of a number of different specifications shows that this is not necessarily the case: with some specifications it matters, while with others it does not.

Interdependent preferences and habit formation are alternative dynamic specifications, but it is often impossible to distinguish between them on the basis of aggregate or per capita demand behavior. Panel data (i.e., observations on the same individuals or households in successive periods) enable us to distinguish between interdependent preferences and habit formation.

2.1. Individual Demand Functions in the Short Run

In this section we examine the individual demand functions corresponding to alternative specifications of interdependent preferences. For expositional convenience we introduce and discuss them in the context of the LES and then show that they can be applied to a wide variety of demand systems.

In the LES the demand function of individual r for good i in period t is given by

$$(43) \qquad x_{it}^r = h^{rit}(P_t, \mu_t^r) = b_{it}^r - \frac{a_i^r}{p_{it}} \sum p_{kt} b_{kt}^r + \frac{a_i^r}{p_{it}} \mu_t^r.$$

Individuals are identified by superscripts on a, b, x, and μ, but everyone faces the same prices.

We incorporate interdependent preferences, into the LES by postulating that some of the demand system parameters depend on other people's consumption. For reasons of tractability, we assume that the a's are constant and that interdependence operates through the b's. The most straightforward assumption is that the b's depend linearly on other people's consumption

$$(44) \qquad\qquad b_{it}^r = b_i^{r*} + \sum_{\substack{s=1 \\ s \neq r}}^{R} \beta_i^{rs} x_{it}^s,$$

where R is the number of individuals. We expect the interdependence coefficients, the β's, to be nonnegative, so that an increase in someone else's consumption of a good implies, ceteris paribus, an increase in its consumption by individual r. As in habit formation models, dynamic stability implies restrictions on the β's; we discuss these restrictions in Section 2.4. As a specification of interdependent preferences, (44) has two major defects. First, it involves too many interdependence coefficients— $n \times (R - 1)$ for each individual. Second, because each person's tastes depend on everyone else's current consumption, the simultaneous determination of an equilibrium consumption pattern for everyone in the society is a formidable task.

Instead of specifying that each individual's preferences depend on everyone else's current consumption, we shall assume that they depend on other people's past consumption. This assumption has the merit of analytical tractability, a virtue not to be despised. It is also consistent with the plausible belief that the acquisition of preferences is an integral part of an ongoing process of socialization. It is tempting to argue that lagged interdependence is more plausible than simultaneous interdependence, but because we have not specified the length of the time periods, such an argument would be fragile.

To incorporate lagged interdependence into the LES, we postulate that b_{it}^r depends linearly on every individual's consumption of good i in the previous period:

$$(45) \qquad\qquad b_{it}^r = b_i^{r*} + \sum_{s=1}^{R} \beta_i^{rs} x_{it-1}^s.$$

This specification differs from (44) both because it embodies the hypothesis of *lagged* rather than simultaneous interdependence, and because the summation in (45) runs over *all* individuals, not just *all other* individuals. One might argue that pure interdependence requires the "own effect" β_i^{rr} to be 0, and that when it is nonzero the interdependence model is contaminated by traces of habit formation; nevertheless, we prefer the more general model, especially since we are free to consider pure interdependence as a special case.

The linear interdependence specification (45) involves an unmanageably large number of interdependence coefficients or "weights": unless there is a systematic relationship among the β's, each individual assigns $n \times R$ independent weights. We now consider alternative assumptions that reduce the interdependence parameters to a manageable number.

We begin by treating two closely related cases. In the first, each individual gives equal weight to everyone's past consumption, including his or her own:

$$(46) \qquad \beta_i^{rs} = \beta_i^r, \qquad s = 1, \ldots, R.$$

In the second, each individual gives equal weight to everyone else's past consumption and 0 weight to his or her own:

$$(47) \qquad \beta_i^{rs} = \begin{cases} \beta_i^r, & s \neq r \\ 0, & s = r \end{cases}.$$

Under (46), (45) becomes

$$(48) \qquad b_{it}^r = b_i^{r*} + \hat{\beta}_i^r \bar{x}_{it-1},$$

where

$$(49) \qquad \bar{x}_{it-1} = \frac{1}{R} \sum_{s=1}^{R} x_{it-1}^s \quad \text{and} \quad \hat{\beta}_i^r = R\beta_i^r.$$

Under (47) it becomes

$$(50) \qquad b_{it}^r = b_i^{r*} + \hat{\beta}_i^r \bar{x}_{it-1} - \beta_i^r x_{it-1}^r.$$

Both of these cases enable us to discuss interdependent preferences in terms of per capita past consumption, $\bar{x}_{it-1}$, although in (50) a further adjustment is made for the individual's own past consumption.

Substituting (48) and (50) into the LES and dropping individual subscripts on a and time subscripts on p and μ yields the demand functions of individual r:

$$(51) \qquad x_{it}^r = b_i^{r*} - \frac{a_i}{p_i} \sum p_k b_k^{r*} + \frac{a_i}{p_i} \mu + \hat{\beta}_i^r \bar{x}_{it-1} - \frac{a_i}{p_i} \sum p_k \hat{\beta}_k^r \bar{x}_{kt-1}$$

$$(52) \qquad x_{it}^r = b_i^{r*} - \frac{a_i}{p_i} \sum p_k b_k^{r*} + \frac{a_i}{p_i} \mu + \hat{\beta}_i^r \bar{x}_{it-1} - \frac{a_i}{p_i} \sum p_k \hat{\beta}_k^r \bar{x}_{kt-1}$$

$$- \beta_i^r x_{it-1}^r + \frac{a_i}{p_i} \sum p_k \beta_k^r x_{kt-1}^r,$$

respectively.

This model of interdependent preferences, like the habit model, can be extended in a number of directions. For example, per capita consumption in the more distant past may influence current tastes, or b may depend linearly on the previous peak of per capita consumption, with or without memory coefficients that allow the influence of the peak to diminish as it recedes into the past. The linearity assumption can be dropped to permit the b's to be nonlinear functions of per capita consumption in the previous period, or, more generally, of per capita consumption in the more distant past.

An individual's preferences are likely to be influenced more by the consumption of those with whom he or she has close contact than by those with whom contact is more distant because, as Duesenberry [1949, p. 27] points out, changes in tastes are caused by frequent contact with superior goods, not by mere knowledge of their existence. The per capita lagged consumption specification, which requires that everyone's lagged consumption count equally, is inconsistent with this observation. We now turn to specifications in which individual demand depends on the distribution of consumption in the previous period, not just on its average. The models we consider are based on the premise that individuals are arrayed in a hierarchy in which each individual's preferences are influenced by the consumption behavior of higher ranked individuals. For definiteness, one can imagine that each individual is attempting to imitate the consumption behavior of individuals perceived as having higher social status, but introducing social status or any other new explanatory variable is unnecessary. Position in the hierarchy is defined in terms of who influences whom, and no sociological concepts are required. It is convenient to number individuals in terms of their position in the hierarchy, so that Person 1 is at the top and Person R at the bottom. To simplify the exposition, we assume that there are no "ties," although these could be accommodated without altering the substance of the analysis.

A simple model of interdependent preferences can be built on the assumption that each individual is concerned only with the consumption of the individual one step above in the hierarchy. The model does not require an individual to recognize the entire hierarchy but only to recognize the next ranked individual; the essential hypothesis is that patterns of influence are consistent with an underlying one-dimensional array. To close the model the behavior of the first individual must be specified. Since there is no one above Person 1, it could be argued that the most appropriate specification is constant tastes (i.e., $\beta_i^{ls} = 0$ for all s, so that $b_{it}^1 = b_i^{1*}$). This, however, implies a sharp break between the preferences of the first individual and the preferences of those immediately below. We prefer to assume that Person 1's preferences depend on own past consumption, as in models of habit formation.

A "two-class" model of interdependent preferences in which the members of the lower class (L) emulate the consumption standards of the upper class (U) is an alternative hierarchy model. Again working with the LES, suppose that the necessary basket of an individual in the lower class depends linearly on the average consumption of those in the upper class. We could assume constant tastes (i.e., $b_{it}^r = b_i^{r*}$) for members of the upper class, but we prefer to postulate that their tastes are influenced by the average past consumption of their own class.

Any of these specifications of interdependent preferences can be applied to any demand system locally linear in expenditure by introducing constant terms, if necessary, and allowing them to depend on the consumption of others. Regularity conditions, however, imply restrictions on admissible parameter values. More generally, any of these specifications can be applied to demand systems that are nonlinear in expenditure, provided the system contains n independent parameters, analogous to the b's: $x_t = h^i(P_t, \mu_t; b_t)$. Except in special cases, however, the resulting demand system will be nonlinear in the consumption of others.

2.2. Per Capita Demand Functions in the Short Run

In Section 2.1 we examined alternative specifications of "linear" inter- dependent preferences and discussed their implications for individual short-run demand behavior. We now consider a society made up of individuals whose preferences are interdependent and examine per capita short-run demand behavior. Our emphasis on per capita rather than market demand functions is a matter of form, not substance; we emphasize per capita results to facilitate comparisons with individual behavior under habit formation (Section 1) and with individual short-run behavior under interdependent preferences (Section 2.1).

Although the per capita short-run demand functions are of interest for their own sake, we treat them primarily as a bridge between the individual short-run demand functions and the per capita long-run demand functions we discuss in Section 2.3. We emphasize the per capita long-run demand functions because the ramifications of interdependent preferences are manifested only in the long run.

Although we base our exposition on the LES, the results generalize immediately to any demand system locally linear in expenditure. If the per capita short-run demand functions are locally linear in per capita expenditure, then the distribution of expenditure has no direct effect on the long-run consumption pattern; its influence must be indirect and operate through interdependent preferences. If the per capita short-run demand functions are locally linear in per capita expenditure and independent of the distribution of past consumption, then the long-run demand system will be independent of the distribution of per capita expenditure. We shall see that under some specifications of interdepen- dent preferences, the distribution of past consumption is a significant determinant of the short-run consumption pattern, while under others it is not.

We opened our discussion of interdependent preferences by considering two specifications in which an individual gives equal weight to everyone else's past consumption. They differed in that one specification, (46), gives the individual's own past consumption the same weight as anyone else's, while the other, (47), gives it no weight. With the LES the individual demand functions corresponding to these two specifications are given by (51) and (52), respectively. To examine the per capita demand functions, we assume that everyone has the same marginal budget shares and the same interdependence coefficients, that is,

$$\text{(53)} \qquad\qquad a_i^r = a_i, \qquad r = 1, \ldots, R$$

$$\text{(54)} \qquad\qquad \beta_i^r = \beta_i, \qquad r = 1, \ldots, R.$$

These assumptions enable us to drop the individual superscripts on the a's and β's.

The per capita demand functions are obtained by summing the individual demand functions over all individuals and dividing by R. If each individual gives equal weight to everyone's consumption in the previous period including his own, then (51) yields

$$\text{(55)} \qquad \bar{x}_{it} = \bar{b}_i^* - \frac{a_i}{p_i} \sum p_k \bar{b}_k^* + \frac{a_i}{p_i} \bar{\mu}_t + \hat{\beta}_i \bar{x}_{it-1} - \frac{a_i}{p_i} \sum p_k \hat{\beta}_k \bar{x}_{kt-1},$$

where $\bar{b}_i^*$ and $\bar{\mu}_t$ denote the average values of b_i^{r*} and μ_t^r:

$$\text{(56)} \qquad\qquad \bar{b}_i^* = \frac{1}{R} \sum_{s=1}^{R} b_i^{s*}, \qquad \bar{\mu}_t = \frac{1}{R} \sum_{s=1}^{R} \mu_t^s.$$

If each individual gives equal weight to everyone else's consumption in the previous period and no weight to his own, then (52) yields

$$\text{(57)} \qquad \bar{x}_{it} = \bar{b}_i^* - \frac{a_i}{p_i} \sum p_k \bar{b}_k^* + \frac{a_i}{p_i} \bar{\mu}_t + \beta_i^* \bar{x}_{it-1} - \frac{a_i}{p_i} \sum p_k \beta_k^* \bar{x}_{kt-1},$$

where β_t^* is defined by

$$\text{(58)} \qquad\qquad \beta_i^* = (R-1)\beta_i = \hat{\beta}_i - \beta_i.$$

Thus, the alternative specifications (46) and (47) lead to per capita demand systems that differ only in the *interpretation* of one parameter. It would not be surprising to find that these specifications imply similar per capita demand functions—if the number of individuals is large and income is fairly evenly distributed, then the difference between the average of everyone's past consumption and the average of everyone else's past consumption must be small. But our assertion is stronger: the per capita demand functions implied by these two specifications are identical.

In these cases the distribution of consumption in the previous period has no influence on the current per capita consumption pattern. In (55), when each individual is influenced by the average of everyone's past consumption including his own, this is obvious: since the individual demand functions are independent of the distribution of past consumption, so are the per capita demand functions. In (57), when each individual is influenced by the average of everyone else's past consumption, individual demand is influenced to some degree by the distribution of past consumption, although the influence is small unless the community is small or the distribution of past consumption very unequal; per capita demand, however, depends only on average past consumption, and not on its distribution.

We now examine the implications for the per capita demand functions of the hierarchy models of interdependent preferences in which individuals are influenced by the consumption behavior of persons of higher rank. Again assume that everyone has the same a's and β's. When each individual is concerned with the consumption of the person immediately above in the hierarchy, one might expect the distribution of past consumption to be a major determinant of the per capita consumption pattern. Surprisingly enough, it is not. It is straightforward to show that the only parameter of the distribution of past consumption other than its average that enters the per capita demand functions is the "range" of the distribution, $x_{it-1}^1 - x_{it-1}^R$. (Our use of "range" is nonstandard; x_1 and x_R need not be the highest and lowest consumption in the community.) The range enters only in terms of the form $\bar{x}_{it-1} + (x_{it-1}^1 - x_{it-1}^R)/R$, and average past consumption, $\bar{x}_{it-1}$, is likely to swamp the effect of the range, because the latter is divided by R. Hence, unless the community is small or the distribution of past consumption very unequal, the influence of the distribution of past consumption is small.

In the hierarchy model it is straightforward to show that the distribution of past consumption has a substantial impact on individual consumption patterns but not on the per capita consumption pattern. A similar result holds under habit formation, where each individual's demand depends only on own past consumption, yet the implied per capita demand functions are independent of the distribution of past consumption.

The two-class model provides a simple example in which the distribution of past consumption has a substantial impact on the per capita consumption pattern. In general, if individual demand functions are independent of the distribution of past consumption, then so are the per capita demand functions. The converse, however, need not hold: the per capita demand functions may be independent of the distribution of past consumption even when the individual demand functions are not.

2.3. Long-Run Demand Functions

We now examine the "long-run" demand functions corresponding to various specifications of interdependent preferences. As in habit formation models, a steady-state or long-run equilibrium consumption vector is one that, if it prevailed in period 0, would also prevail in period 1. If prices and each individual's total expenditure remain constant over time, the per capita consumption vector in each subsequent period would also equal the consumption vector of period 0.

Only in the long run can we observe the full effects of interdependent preferences. In the short run the distribution of expenditure among individuals can influence the per capita consumption pattern only if different individuals have different marginal budget shares. When the individual short-run demand functions are locally linear in expenditure and everyone has the same marginal budget shares, then the distribution of expenditure among individuals has no direct influence on the per capita short-run consumption pattern, but the distribution of expenditure can influence the long-run consumption pattern indirectly, through interdependent preferences.

Our exposition thus far has been based on the LES. In this section we use a more general framework, namely, demand functions locally linear in expenditure, to emphasize the similarity of the per capita long-run demand functions under interdependent preferences and the individual long-run demand functions under habit formation.

We begin by discussing specifications of interdependent preferences in which each individual is influenced by average past consumption. There are two such specifications, (46) and (47), which differ only in their treatment of the individual's own past consumption. Because both specifications imply the same per capita short-run demand functions, they must also imply the same per capita long-run demand functions.

The per capita long-run demand functions can be found by solving this system of short-run demand functions under the assumption that $\bar{x}_{it} = \bar{x}_{it-1} = \bar{x}_i$ for all i. Mathematically, this problem is identical to one considered in Pollak [1976b], where it was necessary to solve the system of short-run demand functions implied by habit formation to find the long-run demand functions. There is no need to repeat here the argument used there; we simply assert its conclusion.

Theorem: Suppose that the short-run per capita demand functions are locally linear in per capita expenditure,

(59) $$\bar{x}_{it} = b_{it} + f_i(P_t) + \gamma^i(P_t)[\bar{\mu}_t - f(P_t) - \sum p_{kt} b_{kt}],$$

and b_{it} is given by

(60) $$b_{it} = \bar{b}_i^* + \beta_i \bar{x}_{it-1}.$$

Then the per capita long-run demand functions are given by

(61) $$h^i(P, \bar{\mu}) = B^i(P) - \Gamma^i(P) \sum p_k B^k(P) + \Gamma^i(P)\bar{\mu},$$

where

(62) $$B^i(P) = \frac{\bar{b}_i^* + f_i(P)}{1 - \beta_i}$$

(63)
$$\Gamma^i(P) = \frac{\gamma^i(P)/(1 - \beta_i)}{\sum p_k \gamma^k(P)/(1 - \beta_k)}.$$

Interdependent preferences do not imply that the distribution of expenditure among individuals influences the long-run consumption pattern. With some specifications distribution matters, and with others it does not. Our theorem implies that under the specifications of interdependent preferences in which each individual is influenced by average past consumption, the per capita long-run demand functions depend only on per capita expenditure, not on its distribution.

Now consider the hierarchy specification in which each individual is influenced by the past consumption of the individual ranked immediately above. Assuming only that the demand functions are locally linear in expenditure, these per capita short-run demand functions can be solved for the long-run equilibrium values, just as under habit formation. The per capita long-run demand functions, like their short-run counterparts, depend on per capita expenditure and on the range of the distribution of consumption. It is awkward to include the long-run consumption of individuals 1 and R as arguments of the per capita long-run demand functions. If habit formation determines the long-run consumption pattern of individual 1, then we can eliminate X^1 from the per capita long-run demand functions, replacing it by a function involving μ_1. We cannot, however, eliminate the X^R in a similar manner, so we retain the form involving both X^1 and X^R because it is more transparent.

Except for the terms involving the range, these demand functions are identical to those implied by the average past consumption specification, (61). In the limit, as the size of the community increases while the range of the distribution of consumption remains bounded, the terms involving the range approach 0 and the demand functions approach those of the average past consumption specification. In the limit, the distribution of expenditure among individuals is no longer a determinant of the long-run consumption pattern. One must take care, however, in applying this conclusion to large finite communities, because we would not expect the range to be independent of the size of the community.

Now consider the two-class specifications in which the tastes of both classes depend on the average past consumption of the upper class. Assuming linearity in expenditure and the same interdependence coefficients for both classes, the per capita short-run demand functions are easily determined. Since the behavior of the upper class is determined by habit formation, the per capita long-run demand functions of the upper class are given by the habit formation formula. Substituting this result into the per capita short-run demand function for the lower class yields the long-run consumption pattern of the lower class. It is easily seen that in the two-class model the long-run per capita consumption pattern depends on the distribution of expenditure between the classes.

2.4. Lagged Interdependence, Simultaneous Interdependence, and Dynamics

In this subsection we tentatively evaluate the specifications of lagged interdependent preferences introduced above and compare them with simultaneous interdependence, a specification in which preferences are influenced by other people's consumption. We then discuss the dynamics of interdependent preferences and conclude by describing a specification that combines interdependence and habit formation.

Because models of interdependent preferences begin with assumptions about the preferences or behavior of an individual, they are suitable for interpreting household budget data. We have emphasized their implications for per capita rather than individual demand functions because the full ramifications of interdependent preferences depend on the reciprocal interaction of all individuals. The outcome of this interaction manifests itself only in the long run. We have focused on demand functions locally linear in expenditure, so that the distribution of expenditure has no direct effect on per capita consumption patterns. We have focused on demand functions that depend linearly on other people's consumption to keep the model tractable. Although our initial intuition was that with interdependent preferences the expenditure distribution would be a determinant of the per capita long-run consumption pattern, we found that, for a number of plausible specifications, it is not. Whether the expenditure distribution matters depends on which specification of interdependence is employed; the notion that interdependence automatically implies a role for expenditure distribution is false.

Because economists' views of interdependent preferences have been largely shaped by the simultaneous specification, some will be uncomfortable with our lagged specification. Simultaneous interdependence implies that the complete adjustment from one equilibrium to another takes place in a single period. Because periods are not "instants," it is misleading to characterize simultaneous interdependence as implying instantaneous adjustment, but it does imply that responses to changes in prices and expenditure work themselves out in a single period. If the periods are long enough, say a decade or a generation, full adjustment in a single period is not a problem, but if periods are years or quarters, the gradual adjustment implied by lagged interdependence seems more plausible.

Corresponding to each specification of pure lagged interdependence (i.e., lagged interdependence untainted by habit formation) is a related specification of simultaneous interdependence. If a specification of pure lagged interdependence is replaced by the corresponding model of simultaneous interdependence—that is, by the specification obtained by replacing each lagged value by the corresponding current value—then the demand functions of the simultaneous specification are identical with the long-run demand functions of the lagged specification. With simultaneous interdependence there is no distinction between the short run and the long run, because the long run works itself out in a single period. Thus, lagged interdependence is both more plausible and more tractable than simultaneous interdependence.

The dynamics of interdependent preferences differ from those of habit formation. Even though the per capita demand functions are identical, individual demand functions are not, and this complicates regularity conditions and stability. In the LES, regularity conditions require that each individual's expenditure be at least as great as the cost of the individual's "necessary basket." With habit formation, the necessary basket depends on the individual's own lagged consumption and the demand system is locally stable if the β's are all less than 1. With interdependent preferences, if an individual is influenced by the average of everyone's past consumption, then regularity conditions will be violated for individuals at the bottom of the expenditure distribution unless the β's are substantially less than 1. With interdependent preferences, if an individual is influenced by the consumption of the next ranking individual in the expenditure hierarchy, and if the distribution of expenditure among individuals does not change drastically from one period to the next, then β's slightly less than 1 imply stability.

Habit formation and interdependent preferences can operate together. In particular, suppose that each individual is equally concerned with everyone else's past consumption, so that $\beta_i^{rs} = \beta_i^r$, $r \neq s$, and that β_i^{rr} assumes a different (nonzero) value. The implied individual demand functions are slightly messy, but the per capita short-run demand functions (derived under the assumption that all individuals have the same interdependence coefficients, the same marginal budget shares, and the same habit formation coefficients) are of the same form as those obtained under the separate hypotheses of habit formation and interdependent preferences. It follows that the per capita long-run demand functions are of the same form as the long-run demand functions under either hypothesis alone. Hence, on the basis of the per capita demand behavior, one cannot distinguish among habit formation, interdependent preferences, and a combination of the two: distinguishing requires data on individual behavior. Thus, although the implications of interdependent preferences and habit formation happen to coincide for one particular type of data, they are empirically distinct and distinguishable hypotheses.

3. WELFARE

Taste formation and change pose difficult problems for welfare analysis. Variable tastes undermine the normative significance of the fundamental theorem of welfare economics, which asserts—in a precise sense and under fairly stringent assumptions—that, in competitive equilibrium, people get what they want, subject to the constraints imposed by technology, resources, and the satisfaction of the wants of others. If tastes are sufficiently malleable, however, this theorem may be no more than a corollary of the more general proposition that people come to want what they get.

Taste differences among households associated with differences in their demographic characteristics have long been a mainstay of the analysis of household budget data, but specifications of taste formation and taste change have only recently come to play a role in the analysis of time series data. The economist's traditional reluctance to investigate taste change is well expressed by Milton Friedman [1962]. After discussing the relative nature of human wants, he writes

> Despite these qualifications, economic theory proceeds largely to take wants as fixed. This is primarily a case of division of labor. The economist has little to say about the formation of wants; this is the province of the psychologist. The economist's task is to trace the consequences of any given set of wants. The legitimacy of and justification for this abstraction must rest ultimately, in this case as with any other abstraction, on the light that is shed and the power to predict that is yielded by the abstraction. (p. 13)

Although Friedman expresses the dominant view, three points should be noted. First, his strictures apply to taste formation and taste change, not to taste differences: thus, the use of demographic characteristics in the analysis of household budget data does not violate his dictum. Second, Friedman recognizes that tastes are not really fixed and, by implication, that they are endogenous to the socioeconomic system. Nevertheless, he is willing to forgo descriptive accuracy, arguing that taste formation and change are not the economist's business and that their study should be left to the psychologist. He regards the proper test of the validity of this intellectual division of labor as its power to predict. This is an essential point, because the recent impetus to treat taste formation and change has come largely from empirical demand analysis. Finally, Friedman's argument presupposes an exclusive concern with "positive economics": whether this division of labor between economics on the one hand and psychology and sociology on the other is appropriate for welfare analysis is a distinct issue that Friedman does not address.

In "De Gustibus Non Est Disputandum," Stigler and Becker [1977] appear to take the more extreme position that economic analysis should shun not only taste formation and change, but also taste differences. They rightly object to invoking taste differences and taste change as a deus ex machina to "explain" whatever we cannot otherwise explain; but they do not take an equally critical view of attributing observed differences or changes in behavior to unobserved differences or changes in household technology. Whether one accounts for the observation that "exposure to good music increases the subsequent demand for good music" (p. 78) in terms of taste change (as we would) or in terms of the accumulation of "music capital" (as Stigler and Becker do) is a matter of semantics, not substance. There is no more explanatory power in a model of household production that postulates the accumulation of unobservable "consumption capital" (e.g., music capital, p. 79) than in a model of habit formation.

Demand analysis and welfare analysis require different interpretations of preferences. In demand analysis the objects of choice are vectors of private decision variables, X, and preferences over them depend on a vector of predetermined "state variables," Z. The state variables may include the individual's own lagged consumption, the current or lagged consumption of others, or demographic variables such as family size. We denote the individual's conditional preference ordering by $R(Z)$ and interpret the statement $X^a R(\bar{Z}) X^b$ to mean that X^a is at least as good as X^b when the state variables are given by $\bar{Z}$. Because the individual takes the state variables as fixed when choosing among vectors of private decision variables, conditional preferences and conditional demand functions provide an adequate foundation for demand analysis. Furthermore if the individual takes the state variables as given when making consumption decisions, then we observe only the conditional demand functions and, from them, we can recover only the conditional preference ordering. In demand analysis we need never ask how an individual would choose between alternatives that differ with respect to the state variables.

In welfare analysis, on the other hand, we often seek to compare the individual's well-being in alternative situations that differ with respect to both private decision variables and state variables. For example, we might ask whether the individual is better off in the status quo or in an alternative situation in which the individual's own consumption of every good is 10% higher while everyone else's is 20% higher. This comparison requires a preference ordering that evaluates the state variables as well as the private decision variables; thus, it cannot be based on conditional preferences but requires unconditional preferences. We denote the unconditional preference ordering (i.e., the ordering over both private decision variables and state variables) by R; the statement $(X^a, Z^a) R(X^b, Z^b)$ means that the individual finds (X^a, Z^a) at least as good as (X^b, Z^b). We denote the "unconditional utility function" that represents the unconditional preference ordering by $U(X, Z)$ and the "conditional utility function" that represents the conditional preference ordering by $U(X; Z)$. The semicolon separating X and Z in the conditional utility function signals that the z's are treated as state variables rather than as choice variables. Although the conditional utility function is an increasing transformation of the unconditional utility function, the transformation itself may depend on the vector of state variables. Formally,

(64) $$U(X; Z) = T[U(X, Z), Z]$$

where the transformation $T(\cdot, z_1, \ldots, z_m)$ is increasing in its first argument. The dependence of the transformation on the vector of state variables reflects the impossibility of recovering unconditional preferences from observed demand or other conditional choices.

If unconditional preferences cannot be inferred from market behavior or from any other conditional choices, how might they be recovered? There are two possibilities. First, in some cases we can observe an individual's unconditional choices, even though they are nonmarket choices. For example, with interdependent preferences we might infer unconditional preferences from an individual's support for alternative redistributive tax and transfer programs. Second, and less congenial to the revealed preference tradition of economics, we might ask individuals about their unconditional preferences ("which would you prefer: the status quo or an alternative situation in which your own consumption of every good is 10% higher while everyone else's is 20% higher?"). These are the only alternatives when the state variables are the current or lagged consumption of others, demographic variables, health status, environmental variables (e.g., climatic conditions or pollution), or goods or services provided by the government (e.g., highways, schools, recreational facilities). With habit formation, when the state variables are the individual's own past consumption, the situation is more complex. Two issues set habit formation apart from other specifications involving state variables.

First, von Weizsäcker [1971] has suggested that with habit formation one can finesse the problem of recovering unconditional preferences by basing welfare comparisons on the "long-run utility function." As Pollak [1976b] argues, this is problematic for two reasons. First, when there are more than two goods, the long-run demand system generated by a habit formation model cannot be rationalized by a long-run utility function, except in very special cases. Second, even when the long-run utility function exists, it is an inappropriate welfare criterion. The difficulty is that the long-run utility function, when it exists, is similar to a "community indifference map" rather than an individual's indifference map. Samuelson [1956] shows that, in very special cases, there exists a community indifference map that rationalizes market demand functions; when the community indifference map exists, it is a convenient device for coding information about market demand behavior, but it has no normative or welfare significance. Similarly, the long-run utility function, in those very special cases in which it exists, is merely a convenient device for coding information about long-run demand behavior; it has no normative or welfare significance.

The second issue distinguishing habit formation from other state variable specifications is the relationship between habit formation and intertemporal utility maximization. The threshold distinction is between models of "naive" habit formation and models of "rational" habit formation (Pollak [1975]). With naive habit formation, in each period the consumer chooses a one-period consumption pattern to maximize a one-period utility function. Thus, in the tradition of empirical demand analysis, naive habit formation models ignore intertemporal allocation. The advantage of naive habit models is tractability. The difficulty is that maximizing successive one-period utility functions is rational only if preferences are separable over time—that is, if the marginal rates of substitution involving consumption within each period are independent of quantities consumed in other periods. Although this assumption simplifies the analysis, it requires us to assume that the individual's behavior is based on a false and repeatedly falsified assumption. The naive habit model assumes that individuals fail to recognize the influence of current consumption on their future preferences, despite their repeated experience of past consumption influencing current preferences.

Models of rational habit formation avoid this difficulty but encounter others. With rational habit formation, individuals correctly recognize that their current consumption affects future behavior. In the resulting intertemporal allocation model, the individual recognizes fully the impact of current consumption on future preferences. This recognition avoids the "irrationality" of naive habit formation, but at the cost of replacing separable one-period preferences by a nonseparable intertemporal preference ordering that ranks lifetime consumption plans. Under rational habit formation an individual's current period consumption pattern will depend on the degree to which each good is habit-forming and on anticipated future prices and expenditure levels. In short, the intertemporal model is incompatible with intertemporal separability.

A further complication in rational habit formation models arises from the need to make explicit assumptions about the individul's ability to "commit" or "precommit" to a lifetime consumption plan. At one extreme, if an individual can precommit without cost, and if the individual accepts the primacy of current preferences, then the optimal lifetime consumption plan is one that maximizes the intertemporal utility function reflecting current preferences. Under these assumptions, rational habit formation is equivalent to maximizing a nonseparable intertemporal utility function. At the other extreme, if precommitment is impossible, then the optimal plan is one that takes full account of the effect of current decisions on future preferences. Under this assumption, the optimal plan is the optimal *feasible* plan, where feasibility takes full account of future decisions that are nonoptimal from the perspective of current tastes. Intermediate cases in which precommitment is costly are even more complex. The conflict between current and future preferences disappears only under special assumptions; Strotz [1955–1956], Pollak [1968], and Peleg and Yaari [1973] discuss "myopia" and "consistent planning."

Becker and Murphy [1988] analyze a model of "rational addiction" in which individuals fully and correctly anticipate the effects of their current consumption on future behavior. Whether models of rational addiction or rational habit formation so complicate the analysis that the resulting model ceases to provide a useful framework for empirical demand analysis is an open question.

An alternative approach to welfare evaluation postulates a long-run or steady-state utility function and a lagged adjustment from one long-run equilibrium to another. Only in a narrow class of cases will there exist a short-run utility function that rationalizes behavior while the adjustment is taking place. We find the assumption of short-run maximizing behavior more appealing than the alternative assumption of long-run maximization, but the choice is a matter of aesthetic preference and research strategy.

5

Stochastic Specifications

In this chapter we outline several stochastic specifications that are important in empirical demand analysis and that form the basis of our empirical results in Chapters 6 and 7. In Section 1 we develop what we refer to as our standard stochastic specification for demand systems written in share form: additive, independent (across observations but not goods), normal errors with 0 mean and a constant nondiagonal contemporaneous covariance matrix. In Section 2 we extend this standard model to allow for first order vector autoregressive systems. We pay particular attention to the effects of including the initial observation in the estimation procedure. We consider two procedures that differ in their treatment of the initial observation: the generalized first difference method in which the initial observation is used only in differencing, and an alternative method in which full use is made of the initial observation. Almost all consumer demand studies in the literature that incorporate serial correlation do so using the generalized first difference method. In Section 3 we develop an error components model in the context of a time series of cross sections. This extends our standard specification to one in which disturbances have a "time-specific" component in addition to a general component. We demonstrate that the resulting likelihood function can be written in a computationally tractable form even though it cannot in general be concentrated with respect to any of the disturbance covariance parameters. Finally in Section 4 we investigate the possibility of estimating random coefficients models for the LES, quadratic expenditure system (QES), and basic translog (BTL) models. We show that estimation of such models is computationally tractable for the LES and QES if the appropriate subset of parameters is assumed to be random. However, the translog forms are less tractable, particularly with more than two goods, and their estimation is beyond the scope of this book.

1. THE STANDARD STOCHASTIC SPECIFICATION

We begin with the simplest, and what we will refer to as our "standard," stochastic specification. We write our system of demand equations for observation t in share form as

(1)
$$w_{it} = \omega^i(z_t, \beta) + u_{it}, \qquad \begin{matrix} i = 1, \ldots n \\ t = 1, \ldots T, \end{matrix}$$

where w_{it} is the share of the ith good in total expenditure, z_t is the set of explanatory variables, β the parameters to be estimated, and u_{it} a random disturbance. Denoting the row vector $(u_{1t}, \ldots, u_{nt})$ as $\tilde{u}_t'$ we assume that $E(\tilde{u}_t \tilde{u}_t') = \tilde{\Omega}$ and that $E(\tilde{u}_t \tilde{u}_s) = 0$ for $s \neq t$. That is, we assume that the contemporaneous covariance matrix for the share disturbances is the same for all observations, and that the disturbances are uncorrelated across observations. We relax the latter assumption in Section 2 when we introduce autocorrelation in the context of a time series model.

The constant covariance assumption appears plausible a priori in view of the fact that the dependent variables are shares, and thus bounded by 0 and 1. On the other hand, if one adds disturbance terms to the demand equations in expenditure form, the assumption of a constant covariance matrix is less plausible. In a cross-section context it implies that the variance of the disturbance associated with expenditure on a good will be the same regardless of the level of expenditure, which may vary widely within a sample. In a time series context it implies that increases in per capita consumption over time will not be accompanied by increases in the disturbances' variances. Furthermore this specification implies that if all prices and total expenditure were to increase proportionately, then the variances of the expenditure equation disturbances would remain constant and thus the variances of the disturbances of the demand equations in share form would decrease, even though the predicted shares remained constant. On theoretical grounds we prefer a specification in which the covariance matrix of the disturbances associated with the demand equations in share form is unaffected by proportional changes in all prices and expenditure. An alternative approach is to assume that the covariance matrix of the disturbances associated with the expenditure equations is proportional to the square of expenditure. This is the approach followed, for example, in Wales [1971]. Finally if one adds disturbance terms to the demand equations in quantity form, then the assumption of a constant covariance matrix for the disturbances is entirely inappropriate. Indeed if there are n linearly independent prices in the sample then this assumption implies that the covariance matrix is of rank 0. For a proof of this proposition see, for example, Pollak and Wales [1969].

The fact that expenditure on the n goods exhausts the budget imposes a restriction on $\tilde{\Omega}$. In particular, summing (1) over all goods gives

(2)
$$\sum w_{kt} = \sum \omega^k(z_t, \beta) = 1,$$

implying that $\sum_k u_{kt} = 0$ for each t, and thus the u_{kt}'s cannot be mutually independent. Further for each t we have $E(\tilde{u}_t \tilde{u}_t') = \tilde{\Omega}$ and $\ell' \tilde{u}_t = 0$, where $\ell' = (1, \ldots, 1)$ a vector of n ℓ's. This implies that $\ell' \tilde{\Omega} = 0$, in which case ℓ' is an eigenvector of $\tilde{\Omega}$ with the corresponding eigenvalue equal to 0; thus $\tilde{\Omega}$ is singular.

We assume that $\tilde{u}_t$ has a multivariate normal distribution with mean 0 and covariance matrix $\tilde{\Omega}$ for all t. Due to the singularity of $\tilde{\Omega}$ the density for $\tilde{u}_t$ may be expressed in terms of the density of any $n-1$ of the goods. Barten [1969] proves that the parameter estimates are independent of the choice of good deleted. This is a major advantage of the maximum likelihood procedure over a two-step Zellner-type procedure for which the estimates depend on the choice of good deleted. We arbitrarily drop the nth good and define $u_t' = (u_{1t}, \ldots, u_{n-1,t})$ and the corresponding covariance matrix as $E(u_t u_t') = \Omega$. Under these assumptions the density for u_t is given by

$$(3) \qquad f(u_t) = (2\pi)^{-[(n-1)/2]} |\Omega|^{-1/2} \exp\left(-\frac{u_t' \Omega^{-1} u_t}{2} \right),$$

and the logarithm of the likelihood function for a sample of T (independent) observations is given by

$$(4) \qquad L(\beta, \Omega) = -\frac{(n-1)T}{2} \log 2\pi - \frac{T}{2} \log |\Omega| - \frac{1}{2} \sum_{t=1}^{T} u_t' \Omega^{-1} u_t.$$

For estimation purposes it is convenient to concentrate this likelihood function with respect to the elements of Ω. Following, for example, Rothenberg and Leenders [1964, p. 61], the concentrated logarithm of the likelihood may be expressed as

$$(5) \qquad L(\beta) = -\frac{(n-1)T}{2} (\log 2\pi + 1) - \frac{T}{2} \log |S|,$$

where S is a square matrix of order $n-1$, with the ijth element given by

$$(6) \qquad s_{ij} = \left(\sum_{t=1}^{T} u_{it} u_{jt} \right) \bigg/ T, \qquad i, j = 1, \ldots, n-1.$$

Thus S is just the sample covariance matrix of the residuals for the first $n-1$ goods. Maximizing the likelihood function is equivalent to minimizing the determinant of S, which is a function only of the β's and the data. This minimization may be carried out using standard nonlinear minimization algorithms. All of the empirical results we present in Chapters 6 and 7 are based on maximizing various likelihood functions using an algorithm due to Fletcher [1972] and rely on numerically calculated derivatives. The material in the first three sections of Chapter 6 is based on maximizing a likelihood function of the form given by (5).

2. MODELS WITH FIRST ORDER AUTOCORRELATED ERRORS

In this section we extend the standard stochastic specification of Section 1 by dropping the assumption of independence across observations. We assume that the $\tilde{u}_t$ follow a first order autoregressive process

$$(7) \qquad \tilde{u}_t = \tilde{R} \tilde{u}_{t-1} + \tilde{e}_t, \qquad t = 2, \ldots, T,$$

where the $\tilde{e}_t$ are independent and normally distributed with mean 0 and a constant contemporaneous covariance matrix. Once again the presence of the budget constraint implies restrictions on some of the covariance parameters. It can be shown that only $(n-1)^2$ independent transformations of the $\tilde{R}$ matrix can be identified. Denote the transformed $\tilde{R}$ matrix by R. Then R is obtained from $\tilde{R}$ by first subtracting the last column of $\tilde{R}$ from each of the other columns of $\tilde{R}$, and then deleting the last row and column. Berndt and Savin [1975] discuss this transformation procedure. Thus we have the following first order autoregressive process

$$(8) \qquad\qquad u_t = Ru_{t-1} + e_t, \qquad t = 2,\ldots,T,$$

where u_t and e_t are both $(n-1) \times 1$ vectors, and the e_t are independently normally distributed with 0 mean and constant covariance matrix Ω. We assume that the process is stationary, implying the u_t are normally distributed with mean 0 and contemporaneous covariance matrix θ, which from (8) satisfies

$$(9) \qquad\qquad \theta = R\theta R' + \Omega.$$

We consider first a procedure that we refer to as the "generalized first difference" procedure, a straightforward generalization to a system of equations of the standard first difference procedure used in the single equation context. This generalized first difference procedure premultiplies the system of $n-1$ equations by R, lags once, and subtracts from the original observation. This yields a system of equations with independent disturbances with mean 0 and covariance matrix Ω. Maximum likelihood estimates of this differenced system do not coincide with maximum likelihood estimates based on the original system and sample: the generalized first difference procedure fails to make full use of the first observation, using it only in differencing. However, the generalized first difference procedure offers a major computational advantage: estimation is relatively straightforward because the likelihood function corresponding to the differenced system can be concentrated and maximized in the usual way. On the other hand, maximum likelihood estimates corresponding to the original system and sample are difficult to calculate because there appears to be no way of concentrating the likelihood function.[1]

With this first differencing procedure the log likelihood has the same form as (4) with $u_t - Ru_{t-1}$ replacing u_t, and $T-1$ replacing T. Further since $\Omega = \theta - R\theta R'$ from (9) we can write the log likelihood as a function of θ, β, and R as follows:

---

[1] Beach and MacKinnon [1979] discuss the problem of concentrating the likelihood function.

$$(10) \quad L(\theta, \beta, R) = k_1 - \left(\frac{T-1}{2}\right) \log|\theta - R\theta R'| - \frac{1}{2} \sum_{t=2}^{T} (u_t - Ru_{t-1})'$$

$$\cdot (\theta - R\theta R')^{-1} (u_t - Ru_{t-1})$$

which, as in the case of (4), can be concentrated with respect to the covariance parameters θ to give

$$(11) \qquad L(\beta, R) = k_2 - \left(\frac{T-1}{2}\right) \log \left| \sum_{t=2}^{T} (u_t - Ru_{t-1})(u_t - Ru_{t-1})' \right|,$$

where k_1 and k_2 are constants independent of θ, β, and R.[2]

Maximization of the likelihood given by (11) is straightforward but slightly more complicated than that given by (4), because the likelihood in (11) depends on the additional parameters contained in the R matrix.

A special case of (11) involves a diagonal R matrix. When R is diagonal, the random disturbance associated with any good depends only on the lagged value of the disturbance for that good, but is independent of lagged disturbances associated with other goods. For each good we replace (8) by

$$(12) \qquad u_{it} = \rho_i u_{it-1} + e_{it}, \qquad i = 1, \ldots, n; \qquad t = 2, \ldots, T,$$

where ρ_i is the serial correlation coefficient associated with the ith good. In this case, however, it can be shown that in order to ensure that the estimates are independent of the equation deleted, all the ρ_i must be equal (because of restrictions resulting from the budget constraint). Berndt and Savin [1975] give a proof of this proposition. Thus, introducing a diagonal R matrix results in only one additional parameter to be estimated. This is attractive from a computational point of view but may be too restrictive to model adequately the true autocorrelation process.

We consider next maximum likelihood estimation of the entire system, a procedure that takes full account of the first observation instead of using it merely to calculate first differences. We assume that $u_1 = Se_1$ where S is defined to ensure the stationary of the u process, that is,

$$(13) \qquad\qquad\qquad \theta = S\Omega S'.$$

The log likelihood function for all T observations is then

$$(14) \quad L(\theta, \beta, R) = k_3 - \frac{T}{2} \log|\theta - R\theta R'| + \frac{1}{2} \log(|\theta - R\theta R'|/|\theta|)$$

$$- \frac{1}{2} \left[u_1' \theta^{-1} u_1 + \sum_{t=2}^{T} (u_t - Ru_{t-1})'(\theta - R\theta R')^{-1}(u_t - Ru_{t-1}) \right]$$

---

[2] The matrix whose determinant is taken in (11) differs by the factor $T-1$ from that in (4), thus giving rise to offsetting differences in the constant terms. We have done this to make our presentation consistent with that of Beach and MacKinnon [1979] on which our analysis is based.

where k_3 is a constant. As noted by Beach and MacKinnon, (14) differs from (10) in two respects: it contains the additional term $-(u_1'\theta^{-1}u_1)/2$ and an additional term that constrains $|\theta - R\theta R'|$ and $|\theta|$ to have the same sign $[\frac{1}{2}\log(|\theta - R\theta R'|/|\theta|)]$. The latter condition is equivalent to requiring the error process (8) to be stationary. That is, including the first observation explicitly constrains the elements of R to guarantee stationarity.

The log likelihood function (14) can be concentrated with respect to θ only when R is a diagonal matrix with a common element ρ on the diagonal. But as mentioned above, when R is diagonal and a system of share or expenditure equations is being estimated, then a diagonal R matrix is restricted to one in which all diagonal elements are the same. In this case (14) becomes

$$(15) \qquad L(\beta, \rho) = k_4 + \frac{(n-1)}{2}\log(1-\rho^2) - \frac{T}{2}\log|(1-\rho^2)u_1 u_1'$$

$$+ \sum_2^T (u_t - \rho u_{t-1})(u_t - \rho u_{t-1})'|$$

where k_4 is a constant.

Maximizing (15) is not much more difficult computationally than maximizing the log likelihood given by (4). However, the unconcentrated log likelihood function given by (14) involves an additional $n(n-1)/2$ unknown parameters as compared with the log likelihood given by (4), and this may represent a significant additional computational burden, especially when the number of goods in the system is large and the share equations are highly nonlinear.

Finally maximization of (14) requires that θ be positive definite. This restriction may be imposed by writing θ as the product of a lower triangular matrix θL and its transpose, and estimating the elements of θL rather than θ.[3]

The empirical results in Chapter 7 are based on maximizing likelihood functions of the form given by (11), (14), and (15).

3. A MODEL WITH TWO ERROR COMPONENTS

When working with a time series of cross sections, a natural extension of the standard stochastic specification is one in which the errors have more than one component. We consider here the case of two error components. Let u_{rt} denote the $(n-1) \times 1$ vector of disturbances added to $n-1$ share equations in a demand system, where, for example, r indexes a household

---

[3] Diewert and Wales [1987] discuss this technique in the context of imposing concavity on cost functions.

in a cross section at time t. These disturbances are assumed to be the sum of two components

$$(16) \qquad u_{rt} = e_t + \varepsilon_{rt}, \qquad \begin{matrix} r = 1, \ldots, q_t \\ t = 1, \ldots, T, \end{matrix}$$

where q_t is the number of households in the period t cross section and T is the number of cross sections. Note that if $r = 1$ for all t then this model reduces to the one discussed in Section 1 and contains T observations. But if $r > 1$ in one or more time periods then the total number of observations will be $\sum_{t=1}^{T} q_t$, which exceeds T. The subscript r does not necessarily denote the same household in successive time periods, nor are there necessarily the same number of households in every period. The $(n-1) \times 1$ disturbance vectors e_t and ε_{rt} are assumed to be independently normally distributed with 0 means and covariances given by Γ and Δ, respectively, where Γ is positive semidefinite and Δ is positive definite. The covariance matrix for the u's is thus

$$(17) \qquad E(u_{rt}u_{s\tau}) = \begin{cases} \Gamma + \Delta & r = s, \quad t = \tau \\ \Gamma & r \neq s, \quad t = \tau \\ 0 & t \neq \tau. \end{cases}$$

In order to write out the likelihood for this model we define

$$(18) \qquad V_t = (u_{1t}, \ldots, u_{q_t t}),$$

which is an $(n-1) \times q_t$ matrix, and

$$(19) \qquad v_t = \text{vec } V_t,$$

which is an $(n-1)q_t \times 1$ vector. Then using (17) we have

$$(20) \qquad \Omega_t \equiv E(v_t v_t') = \begin{cases} S_t S_t' \otimes \Gamma + I_t \otimes \Delta, & t = \tau \\ 0 & t \neq \tau, \end{cases}$$

where $S_t' = (1, \ldots, 1)$ is a $1 \times q_t$ vector of 1's, and I_t is the identity matrix of order q_t. This error components model is now in essentially standard form, and thus the log likelihood function for a sample of T independent time periods is given by (aside from an additive constant)

$$(21) \qquad L(\beta, \Gamma, \Delta) = -\frac{1}{2} \left\{ \sum_{t=1}^{T} \log |\Omega_t| + \sum v_t' \Omega_t^{-1} v_t \right\}$$

with Ω given by (20) and v_t by (19). This likelihood function can now be maximized with respect to the elements of β, Γ, Δ. For estimation purposes it would be convenient if this expression could be concentrated with respect to the Γ and Δ parameters. Although this is possible for the case when the q_t's are equal for all t, it does not appear to be possible when the q_t's differ.

Since Γ and Δ are $(n-1) \times (n-1)$ symmetric matrices our stochastic specification contains $n(n-1)$ independent parameters in addition to the parameter set β. For large n and q_t, (21) may involve a prohibitive number of parameters to be estimated in view of the fact that Ω_t is of order $(n-1)q_t$. Fortunately (21) can be written in a simple form involving matrices of order $n-1$. We may rewrite Ω_t as

$$(22) \qquad \Omega_t = \frac{S_t S_t'}{q_t} \otimes W_t + \left(I - \frac{S_t S_t'}{q_t}\right) \otimes \Delta,$$

where $W_t = \Delta + q_t\Gamma$, in which case it can be shown that[4]

$$(23) \qquad \Omega_t^{-1} = \frac{S_t S_t'}{q_t} \otimes W_t^{-1} + \left(I - \frac{S_t S_t'}{q_t}\right) \otimes \Delta^{-1}$$

and

$$(24) \qquad |\Omega_t| = |W_t| |\Delta|^{q_t - 1}.$$

Further, from the properties of the trace operator and making use of (23) we may write $v_t'\Omega_t^{-1}v_t$ as

$$(25) \qquad v_t'\Omega_t^{-1}v_t = \mathrm{tr}\left[V_t\frac{S_t S_t'}{q_t}V_t'\right]W_t^{-1} + \mathrm{tr}\left[V_t\left(I - \frac{S_t S_t'}{q_t}\right)V_t'\right]\Delta^{-1}$$

$$= \mathrm{tr}\left[\frac{V_t S_t S_t' V_t'(W_t^{-1} - \Delta^{-1})}{q_t}\right] + \mathrm{tr}(V_t V_t'\Delta^{-1})$$

$$= \frac{S_t'V_t'(W_t^{-1} - \Delta_t^{-1})V_t S_t}{q_t} + \mathrm{tr}(V_t'\Delta^{-1}V_t)$$

$$= \sum_{s=1}^{q_t}\sum_{r=1}^{q_t} u_{st}'\frac{(W_t^{-1} - \Delta^{-1})}{q_t}u_{rt} + \sum_{s=1}^{q_t} u_{st}'\Delta^{-1}u_{st}.$$

Substituting (25) and (23) into (21) yields the following log likelihood function (aside from an additive constant):

$$(26) \quad L(\beta, \Gamma, \Delta) = -\frac{1}{2}\sum_{t=1}^{T}\left\{\log|W_t| + (q_t - 1)\log|\Delta| + \sum_{s=1}^{q_t} u_{st}'\Delta^{-1}u_{st}\right.$$

$$\left. + \sum_{s=1}^{q_t}\sum_{r=1}^{q_t} u_{st}'\frac{(W_t^{-1} - \Delta^{-1})}{q_t}u_{rt}\right\}.$$

The computational advantage of (26) lies in the fact that it contains matrices of order $(n-1)$, whereas (21) contains matrices of order $(n-1)q_t$. Thus even with large cross sections it is feasible to maximize the log likelihood function given by (26). However, since (26) cannot in general be concentrated, the $n(n-1)$ covariance parameters must be estimated

---

[4]A proof appears in Lemma 2.1 in Magnus [1982].

along with the demand system parameters. On the other hand, if $q_t = q$ for all t, that is, if each cross section contains the same number of observations, then the likelihood function can be concentrated with respect to Γ and Δ. In this case maximizing the likelihood is no more difficult computationally than maximizing the likelihood in the standard model of Section 1. For a proof see Magnus [1982, p. 251].

We have focused here on a model with two error components because our empirical work in Section 4 of Chapter 6 involves estimating such a model. If the cross section time series data are in panel form, then a model with three error components could be estimated. These components would reflect time-specific and household-specific effects in addition to the standard additive disturbance. Such a model appears, for example, in Avery [1977] and in Baltagi [1980].

4. RANDOM COEFFICIENTS MODELS

In this section we explore the possibility of incorporating a random coefficients stochastic structure into demand systems. In theory any subset of parameters can be stochastic; in practice, unless the stochastic subset of parameters is carefully chosen, the resulting demand system will be computationally intractable. Furthermore, each system must be studied separately to determine its tractability. We consider first the QES and LES models, and then the translog models.

4.1. Quadratic Expenditure System

The nonstochastic Σ-QES demand equations in share form (after deleting the observation subscript) are

$$(27) \qquad w_i = \frac{p_i b_i}{\mu} + a_i \left(1 - \frac{\sum p_k b_k}{\mu} \right)$$

$$+ \left(\frac{p_i}{\mu} c_i - a_i \sum \frac{p_k}{\mu} c_k \right) \prod \left(\frac{p_k}{\mu} \right)^{-2a_k} \left(1 - \sum \frac{p_k}{\mu} b_k \right)^2 .$$

Suppose we now consider replacing a set of parameters θ_i (such as the a_i's, b_i's or c_i's) by $\theta_i + \varepsilon_i$, where ε_i is a normal random variable with mean 0. If this is done for the a's or b's in (27) the resulting error term is computationally intractable and, in particular, it is no longer normal. However, if we replace c_i in (27) by $c_i + \varepsilon_i$, then the resulting disturbance in (27), defined as v_i, is[5]

---

[5]Eq. (27) can be thought of as having been derived by maximizing the corresponding stochastic QES, where the ε_i's are known by the individual, but not by the researcher. The estimated β's are then the mean β's in the population. However, this is neither necessary nor especially useful. We believe that the specification of the error structure should be judged on its implications for the stochastic demand functions.

$$(28) \qquad v_i = \left(\frac{p_i}{\mu} \varepsilon_i - a_i \sum \frac{p_k}{\mu} \varepsilon_k \right) \delta(P, \mu), \qquad i = 1, \ldots, n$$

where

$$\delta(P, \mu) \equiv \prod \left(\frac{p_k}{\mu} \right)^{-2a_k} \left(1 - \sum \frac{p_k b_k}{\mu} \right)^2.$$

Since the v_i's are linear transformations of the ε_i's they are also normally distributed. The v_i's have 0 mean and covariance matrix Ω with ijth element

$$(29) \qquad \Omega_{ij} \equiv E(v_i v_j) = \frac{\delta^2(P, \mu)}{\mu^2} \left[p_i p_j E(\varepsilon_i \varepsilon_j) - p_i a_j \sum p_k E(\varepsilon_i \varepsilon_k) \right.$$

$$\left. - a_i p_j \sum p_k E(\varepsilon_k \varepsilon_j) + a_i a_j \sum \sum_\ell p_k p_\ell E(\varepsilon_k \varepsilon_\ell) \right].$$

Further $\sum v_k = 0$ and we have the errors summing to 0 automatically.

Assumptions about $E(\varepsilon_i \varepsilon_j)$ will be based on their implications for Ω, the covariance matrix associated with the share equation disturbances. First we require that Ω be homogeneous of degree 0 in prices and expenditure. If all prices and expenditure increase proportionately, then the nonstochastic shares remain constant and it seems reasonable to assume the error variances do also. Second we assume that the share equation disturbances are homoskedastic with respect to expenditure. Although the homoskedasticity assumption is not necessary, it seems reasonable a priori, and implies an error structure that closely parallels that studied in Section 1 in which the error covariance matrix is independent of expenditure and prices. In fact we shortly relax this homoskedasticity assumption.

Perhaps the simplest way to obtain a covariance matrix for the share equation disturbances with these properties is to assume that the ε's satisfy

$$(30) \qquad E(\varepsilon_i \varepsilon_j) = \frac{\sigma_{ij} \mu^2}{p_i p_j \delta^2(P, \mu)}, \qquad i, j = 1, \ldots, n,$$

where the σ_{ij} are constant over observations. The resulting ijth element of Ω is then given by

$$(31) \qquad \Omega_{ij} \equiv E(v_i v_j)$$

$$= \frac{p_i p_j \sigma_{ij} - p_i a_j \sum p_k \sigma_{ki} - a_i p_j \sum p_k \sigma_{kj} + a_i a_j \sum \sum_\ell p_k p_\ell \sigma_{k\ell}}{p_i p_j}.$$

Thus Ω is homogeneous of degree 0 in prices and expenditure together and, because it is independent of expenditure, it is also homogeneous of degree 0 in prices alone. These properties are attractive from a theoretical point of view, because the dependent variables are shares. From a computational point of view this error structure is less tractable than the standard formulation of Section 1 because the elements of Ω depend on the sample observations through the prices; hence the likelihood function cannot be concentrated with respect to the σ's.

Because the v's sum to 0 automatically, the budget constraint imposes no additional restrictions on the ε's. Thus, this is a richer specification than the standard stochastic specification in the sense that there are $n(n + 1)/2$ independent covariance parameters (σ's), whereas the standard specification contains $n(n - 1)/2$, yielding a difference of n between the two models. A special case of the random coefficients model is one in which $\sigma_{ij} = 0$ for $i \neq j$, resulting in n independent variances. With $n = 3$ this model contains the same number of covariances as does the standard stochastic specification, and for $n > 3$ it contains fewer. Because the likelihood function cannot be concentrated in the random coefficients model, this restriction may be attractive on computational grounds. Of course in theory the random coefficients and the standard models may be used simultaneously. In this case if the c_i in (27) are replaced by $c_i + \varepsilon_i$, and an additional error term u_i, independent of ε_i, is appended to (27), then the ijth element of the disturbance covariance matrix for the system will be given by (31) plus a constant term (corresponding to the covariances of the u's). This covariance matrix contains $n(n + 1)/2 + n(n - 1)/2 = n^2$ independent elements, and in empirical applications it may well be the case that the data cannot distinguish this many distinct covariance parameters.

We can easily relax the assumption that the share equation disturbances are homoskedastic with respect to expenditure while maintaining the assumption of homogeneity of degree 0 in expenditure and prices. A particularly simple method and one that introduces only one additional parameter (α) is to replace (30) by

$$(32) \qquad E(\varepsilon_i \varepsilon_j) = \frac{\sigma_{ij} \mu^{2\alpha}}{p_i^\alpha p_j^\alpha \delta^2(P, \mu)}, \qquad i, j = 1, \ldots, n.$$

Two special cases of (32) are of interest. If $\alpha = 1$ Eq. (32) reduces to (30) and we have the model just considered, in which share equation disturbances are homoskedastic with respect to expenditure. If $\alpha = 0$ then instead of (31) we have for the share equation disturbances, denoted by v_i^*

$$(33) \qquad E(v_i^* v_j^*) = \frac{p_i p_j}{\mu^2} \Omega_{ij},$$

where Ω_{ij} is given by (31). Since v_i^* is a share equation disturbance, (32) implies that the expenditure equation disturbances ($v_i^* \mu$) are homoskedastic with respect to expenditure. Given an estimate of α in the more general model, these two special cases can be readily tested using standard procedures such as a likelihood ratio test or an asymptotic t test. Of course, more general forms of homoskedasticity could be hypothesized. However, the form given here is parsimonious in terms of additional parameters to be estimated, maintains homogeneity of degree 0 in prices and expenditure, and contains the two special cases, of interest described above.

4.2. Linear Expenditure System

The LES demand equations in share form (after deleting the observation subscript) are

$$(34) \qquad w_i = \frac{p_i b_i}{\mu} + a_i \left(1 - \frac{\sum p_k b_k}{\mu} \right).$$

For this model allowing either the a's or the b's to be stochastic leads to computationally tractable disturbance terms in (34).

Consider first replacing the a_i in (34) by $a_i + \varepsilon_i$ where, as above, the ε_i's are normally distributed random variables with mean 0. The resulting disturbance in (34) defined as v_i, is

$$(35) \qquad v_i = \varepsilon_i \left(1 - \frac{\sum p_k b_k}{\mu} \right), \qquad i = 1, \dots, n.$$

The v_i's have mean 0 and covariance matrix Ω given by

$$(36) \qquad \Omega_{ij} \equiv E(v_i v_j) = E(\varepsilon_i \varepsilon_j) \left(1 - \frac{\sum p_k b_k}{\mu} \right)^2.$$

If we assume that $E(\varepsilon_i \varepsilon_j) = \sigma_{ij}$ is a constant, then this is simply the standard model with heteroskedasticity. Such a model can easily be estimated, but if we wish to impose the condition that the share disturbances be homoskedastic with respect to expenditure, then we could assume that

$$(37) \qquad E(\varepsilon_i \varepsilon_j) = \left[\frac{\mu}{\mu - \sum b_k p_k} \right]^2 \sigma_{ij},$$

in which case $E(v_i v_j) = \sigma_{ij}$, which is the same as the standard stochastic specification.

If we replace the b_i's in (34) by $b_i + \varepsilon_i$, then the disturbance term in (34) becomes

$$(38) \qquad v_i = \frac{p_i}{\mu} \varepsilon_i - a_i \sum \frac{p_k \varepsilon_k}{\mu}, \qquad i = 1, \dots, n.$$

This differs from the corresponding expression for the QES, given by (28), only by the factor δ; hence, $\sum v_k = 0$ automatically. For reasons analogous to those given above for the QES, we assume that the u's satisfy

$$
(39) \qquad\qquad E(\varepsilon_i \varepsilon_j) = \frac{\mu^2}{p_i p_j} \sigma_{ij}.
$$

Because the b's enter the LES in essentially the same way that the c's enter the QES, this specification implies the same Ω matrix for the v's that appears in (31). After adjusting for heteroskedasticity and ensuring that the covariance matrix of share disturbances is homogeneous of degree 0 in expenditure and prices, the covariance matrix of disturbances is the same for the QES with random c's as it is for the LES with random b's.

Denoting this covariance matrix at observation t as Ω_t, the logarithm of the likelihood function for a random sample of T observations (aside from an additive constant) is then given by

$$
(40) \qquad\qquad L(\beta, \Omega) = -\frac{1}{2} \sum_{t=1}^{T} \log |\Omega_t| - \frac{1}{2} \sum_{t=1}^{T} v_t' \Omega_t^{-1} v_t,
$$

where $v_t' = (v_{1t}, \ldots, v_{n-1,t})$ is a $1 \times (n-1)$ vector of disturbances at time t, obtained from either the LES or QES share equations. Because Ω_t depends on t this likelihood cannot be concentrated with respect to the elements of Ω and hence must be maximized with respect to the σ's together with the demand system parameters. In this model, as in the standard one, the estimates are independent of which equation is dropped in the estimation. For a proof see Pollak and Wales [1969].

4.3. Translog Models

The basic translog model in share form (after deleting the observation subscript) is

$$
(41) \qquad w_i = \frac{\alpha_i + \sum_j \beta_{ij} \log(p_j/\mu)}{\sum \alpha_k + \sum \sum_j \beta_{kj} \log(p_j/\mu)}, \qquad \begin{array}{l} \beta_{ij} = \beta_{ji} \quad \text{for all } i, j \\ \sum \alpha_k = 1. \end{array}
$$

Suppose that we replace the α_i in (17) by $\alpha_i + \varepsilon_i$ where, as before, the ε_i are normal random disturbances; then both the numerator and denominator of (41) will be normally distributed. The distribution of a ratio of normally distributed variables was first investigated by Fieller [1932]. More recently Yatchew [1986] has extended Fieller's results to multivariate distributions involving ratios of normal variables. The density function for a set of ratios of normal variables that are not independent, with a common denominator, is in general very complex (Yatchew [1986]). Indeed, the density function has a closed-form representation only when there are an even number of ratios in the set (which, in the context of singular demand systems, implies an odd number of goods), or when there is only one ratio, in which case the density reduces to that derived by Fieller. Yatchew [1985] has estimated a Fieller distribution in the labor supply context, but as far as we know the more general case with more than one independent good has not been estimated, and is beyond the scope of this book.

A tractable special case of this model emerges when, instead of assuming that the ε's are independent, we assume that they satisfy $\sum \varepsilon_k = 0$ for each observation. In this case the ε's do not appear in the denominator and we have

$$(42) \qquad\qquad\qquad E(v_i v_j) = \frac{\sigma_{ij}}{D^2},$$

where D is the denominator in (41). Once again we have the standard model, but with a simple form of heteroskedasticity. To impose the condition that the share disturbances are homoskedastic we assume

$$(43) \qquad\qquad\qquad E(\varepsilon_i \varepsilon_j) = D^2 \sigma_{ij},$$

in which case $E(v_i v_j) = \sigma_{ij}$. Thus in this special case the random coefficient model reduces to the standard stochastic specification.

The empirical results of Section 5 of Chapter 6 are based on maximizing likelihood functions of the form given by (40) for the LES and QES demand systems.

6

Household Budget Data

In this chapter we report demand system estimates based on household budget data[1]. In Section 1 we demonstrate that interesting complete demand systems can be estimated from a small number of budget studies despite the limited price variability represented in such data. To illustrate we use household data for two years to estimate the LES and the QES. In Section 2 using a larger sample we concentrate on functional form specification and demographic influences on consumption patterns. We compare the LES with the quadratic expenditure system (QES), and the basic translog (BTL) with the generalized translog (GTL). We incorporate the age composition and the number of children using linear demographic translating and linear demographic scaling. In Section 3 we investigate three additional methods of incorporating demographic characteristics; because of the complexity of these procedures we estimate a simpler demand system—the generalized CES—rather than the QES or translog forms. In Section 4 we explore two dynamic QES specifications and introduce a stochastic structure that allows disturbances in each time period to be correlated across households. Due to the complexity of the estimation procedure we analyze separately households with different numbers of children. In Section 5 we explore random coefficient models for the LES and QES.

1. ESTIMATION FROM TWO BUDGET STUDIES

Both the LES and QES can be estimated using budget studies for two years. For any demand system, household budget data for a single period identify the income–consumption curve corresponding to the period's prices and, hence, the marginal budget shares at every level of expenditure. We consider first the LES, in which the demand equations in share form are given by

(1) $\quad p_i x_i = p_i b_i + a_i(\mu - \sum p_k b_k), \qquad a_i > 0, \qquad (x_i - b_i) > 0, \qquad \sum a_k = 1.$

---

[1] This chapter is drawn in part from Pollak and Wales [1978, 1980, 1981] and Darrough, Pollak, and Wales [1983].

In the LES, the marginal budget shares are independent of prices and the level of expenditure and are equal to the a's. Thus, household budget data for a single period identify the a's. If one of the b's is known a priori, then budget data for a single period are enough to identify also the (n-1) remaining b's.[2] Even if none of the b's is known a priori, budget data for two periods identify all of the parameters of the LES: data from each period identify the corresponding income–consumption curve, and the intersection of the two linear income–consumption curves uniquely determines the point $(b_1, \ldots, b_n)$.

For the Σ-QES the demand equations in expenditure form are given by

$$(2) \quad p_i x_i = p_i b_i + a_i(\mu - \sum p_k b_k) + (p_i c_i - a_i \sum p_k c_k) \prod p_k^{-2a_k}(\mu - \sum p_k b_k)^2,$$

$$\sum a_k = 1.$$

We now prefer this form to the λ-QES, which we estimated in Pollak and Wales [1978]. Budget data for two periods identify all of the parameters of the QES. The following heuristic argument, although not formally decisive, indicates why.[3] Data from each period identify the income–consumption curve corresponding to that period's prices, and since all income–consumption curves radiate upward from the point $(b_1, \ldots, b_n)$, the intersection of two income–consumption curves determines the point $(b_1, \ldots, b_n)$. Estimates of the b's enable us to calculate supernumerary expenditure $(\mu - \sum p_k b_k)$ for each household in each period. The a's are the coefficients of supernumerary expenditure, while the c's can be disentangled from the coefficients of $(\mu - \sum p_k b_k)^2$.

In contrast to the LES and QES, which can be estimated from two cross sections, in the translog demand system the number of budget studies needed to estimate all of the parameters varies with the number of goods (n).[4] In particular, it can be shown that M cross sections provide enough information to permit estimation of a translog system with at most $n + 1 + M(n-1)$ parameters, while the nonhomothetic basic translog system contains $(n^2 + 3n - 2)/2$ independent parameters. Hence, two cross sections are sufficient to identify the translog with two or three goods, but not the translog with four or more goods; three cross sections identify it with four goods, but not with five or more.

---

[2] Howard Howe [1975] has shown that the identification of the parameters of Lluch's [1973] extended LES can be interpreted in precisely these terms: Lluch's specification implicitly assumes that the b associated with saving is 0.

[3] More formally, suppose that we rewrite (2) as $p_i x_i = \theta_{1i} + \theta_{2i}\mu + \theta_{3i}\mu^2$ and estimate the θ's for $n - 1$ goods in both periods. Then it is easy to show that after substituting the θ_{3i} values into the θ_{2i} values we can obtain estimates of the a's and of $\sum p_k b_k$ for both periods. Using these we can then obtain estimates of the individual b's from the θ_{1i} values, and estimates of the c's from the θ_{3i} values.

[4] See Lau, Lin, and Yotopoulos [1978] for estimates of a translog for Taiwan based on household budget data.

The consumption data used in this chapter are obtained from the *Family Expenditure Survey* series (U.K. Department of Employment and Productivity), an annual publication that reports expenditure patterns of households in the United Kingdom, cross-classified by income and family size.[5] In this section we use data for the two years 1966 and 1972; to simplify computations, we analyze only three consumption categories, "food," "clothing," and "miscellaneous."[6] For 1966 we have mean expenditure on each consumption category by families in four income classes and three family size classes: "one child," "two children," and "three or more children." For 1972, we have six income classes for families with one child and two children, five income classes for families with three children, and three income classes for families with "more than three children." These 32 cells form our basic data and we treat them as if they represent the consumption patterns of households rather than cell means.[7] Retail price indexes corresponding to these categories are taken from the *Annual Abstract of Statistics*, 1974.

We obtain a stochastic form for the LES and QES by adding a disturbance term to the share form of each demand equation. We use the share forms because they are likely to involve less heteroskedasticity than the expenditure forms. We demote the 3×1 vector of disturbances corresponding to the ith cell by $u_i = (u_{i1}, u_{i2}, u_{i3})$ and assume that $E(u_i) = 0$, that $E(u_i u_i') = \Omega$ for all i, and that the u_i's are independently normally distributed. This is the standard model discussed in Section 1 of Chapter 5; hence, the concentrated log likelihood is given by Eq. (3) in Chapter 5.

Estimates of the LES and the QES obtained in this way satisfy the regularity conditions required by economic theory. The Slutsky matrix implied by our parameter estimates is negative semidefinite at each of the

---

[5] A detailed discussion of the data used in this chapter appears in Appendix A.

[6] Since the purpose of this section is to illustrate demand estimation with as few as two cross sections, it is of little importance which years are selected provided they are far enough apart to provide some price variability. We choose 1966 and 1972 because we estimate systems that include the number of children as a determinant of consumption behavior and the period 1966–1972 is the longest period over which "children" were defined consistently. In the surveys before 1966 the number of persons rather than the number of children was reported. After 1972 persons were classified as children if they were 18 or under, while in earlier years they were so classified if they were 16 or under.

[7] Since our consumption data are cell means rather than observations on individual families, an aggregation problem arises. With the LES, it causes no difficulty because the mean consumption pattern of a group is the consumption pattern corresponding to the group's mean expenditure. But with the QES, the mean consumption pattern of a group depends on the variance as well as the mean of the group's expenditure. Unfortunately, we do not know these variances, so, faute de mieux, we have treated the reported cell means as if they were observations on individual families. Aggregation of demand systems quadratic in the independent variable is discussed by Klein [1962, pp. 25–26] and Diewert [1974b, pp. 129–130].

Table 1 Marginal Budget Shares and Own-Price Elasticities: LES and QES

Expenditure (S. per week)	Consumption category		
	Food	Clothing	Miscellaneous
Marginal budget shares: QES			
200	.48	.25	.27
300	.39	.21	.40
400	.31	.17	.52
Own-price elasticities: QES			
200	−.56	−.53	−.41
300	−.76	−1.16	−.90
400	−1.01	−1.80	−1.25
Own-price elasticities: LES			
200	−.62	−1.06	−1.35
300	−.71	−1.04	−1.17
400	−.76	−1.02	−1.11

Notes:
1. Mean expenditure in 1970 is 315 S. per week.
2. For the LES the marginal budget shares are independent of expenditure and are .35, .20, and .45.

32 price expenditure situations corresponding to our data.[8] Further, the QES represents a significant improvement over the LES according to the usual likelihood ratio test; the 5% critical value is 7.8, while the value of $-2\log\lambda$ (where λ is the ratio of the likelihoods associated with the LES and the QES) is 10.4.

Perhaps the most interesting comparison to be made between the LES and the QES involves the behavior of the marginal budget shares. For the LES they are the estimated a_i parameters, while for the QES they depend on all of the estimated parameters. In Table 1 we present the marginal budget shares corresponding to 1970 prices for three expenditure levels.[9] For the QES the shares vary considerably with expenditure levels: at the 200 shilling expenditure level 48% of an additional shilling goes to food and at the 400 shilling level 31% of an additional shilling goes to food. For the LES the share is 35% at all expenditure levels.

---

[8] Our approach to the evaluation of regularity conditions differs from that adopted by Jorgenson and Lau [1979] in their discussion of the translog. They consider only whether regularity conditions are satisfied at a single point, the point of approximation of the translog, whereas we have checked them at every sample point. On the other hand, Jorgenson and Lau test the significance of the ability of their estimated translog to satisfy the regularity conditions, whereas we do not.

[9] This choice of year and expenditure levels makes possible comparison of these results with those of succeeding sections in this chapter.

Corresponding patterns for clothing and miscellaneous can be read from the table The own-price elasticities are also presented in Table 1 and range from $-.41$ to -1.80 for the QES, and from $-.62$ to -1.35 for the LES. In general the QES elasticities vary more with expenditure levels than do the LES elasticities, and the two sets of estimates suggest a different pattern of price responsiveness for the clothing category.

In summary we have estimated two complete demand systems—the LES and the QES—using U.K. household budget data for two periods. The empirical results are generally good in that the regularity conditions are satisfied at all sample points and the price and income responses appear to be reasonable, a priori.

The QES is well-suited to the analysis of data sets with severely limited price variation. With household budget data, each study identifies an income–consumption curve, and, although none of the underlying parameters of the QES can be identified from a single income–consumption curve, all of the parameters can be identified from two such curves.

2. FUNCTIONAL FORM COMPARISONS

We now use a time series of five budget studies to analyze the effects of demographic variables on expenditure patterns and to study the suitability of several different functional forms. We use the period 1968–1972 because this is the longest period over which the demographic variables are defined consistently.

We use the same consumption categories as in Section 1: food, clothing, and miscellaneous. The Family Expenditure Survey cross-classifies households by income and number of children, and for each cell it reports mean expenditure on the three consumption categories and the mean number of children under 2, the mean number under 5, and the mean number under 16. These 81 cells are our basic data and we treat them as if they represent the consumption patterns of households rather than cell means.[10]

The demand systems that we estimate are the basic translog (BTL), the generalized translog (GTL), and the quadratic expenditure system (QES).[11] In share form the demand equations for the BTL are

$$(3) \quad w_i = \frac{\alpha_i + \sum_j \beta_{ij} \log(p_j/\mu)}{\sum \alpha_k + \sum \sum_j \beta_{kj} \log(p_j/\mu)}, \qquad \beta_{ij} = \beta_{ji}, \qquad \sum \alpha_k + \sum \sum \beta_{kj} = 1,$$

---

[10] We describe the composition of these cells in Appendix A.

[11] We also estimated the LES and linear translog (LTL). However, likelihood ratio tests indicate that the QES is superior to the LES at the 1% level and that the GTL is superior to both the LES and the LTL at the 1% level.

where the α's and the β's are parameters to be estimated.[12] The GTL demand equations, in share form, are given by

$$(4) \quad w_i = \frac{b_i p_i}{\mu} + [1 - (\sum p_k b_k)/\mu] \frac{\alpha_i + \sum_j \beta_{ij} \log[p_j/(\mu - \sum p_k b_k)]}{\sum \alpha_k + \sum \sum_j \beta_{kj} \log[p_j/(\mu - \sum p_k b_k)]},$$

where the α's, β's and b's are parameters to be estimated.

The Σ-QES demand equations in share form are given by

$$(5)$$

$$w_i = \frac{p_i b_i}{\mu} + \alpha_i \left(1 - \frac{\sum p_k b_k}{\mu}\right) + \left(\frac{p_i c_i}{\mu} - \alpha_i \sum \frac{p_k c_k}{\mu}\right) \prod \left(\frac{p_k}{\mu}\right)^{-2\alpha_k} \left(1 - \frac{\sum p_k b_k}{\mu}\right)^2,$$

$$\sum \alpha_k = 1.$$

We consider both linear demographic translating and linear demographic scaling as alternative procedures for introducing demographics into these demand systems. In the QES and GTL, introducing demographic variables by linear translating is equivalent to specifying that the translation parameters in these systems (the b's) are linear functions of the demographic variables.[13] The BTL contains no translation parameters; rather than apply demographic translating to a demand system without such constant terms, we modify the BTL by introducing them. Since this yields the GTL, our discussion of demographic translating involves only two distinct demand systems rather than three. Introducing demographic variables by linear scaling requires that each p_i on the right-hand side of the share equations be replaced by $p_i m_i$, where m_i is a linear function of the demographic variables.[14] For each cell, our data report the average number of children per household in each of three age categories: η_2, children under 2 years of age; η_3, children under 5 but at least 2 years of age; and η_4, children 5 and over but under 17. Hence for scaling we have $m_i = 1 + \bar{\varepsilon}_{2i}\eta_2 + \bar{\varepsilon}_{3i}\eta_3 + \bar{\varepsilon}_{4i}\eta_4$ and for translating we have $b_i = b_i^* + \bar{\delta}_{2i}\eta_2 + \bar{\delta}_{3i}\eta_3 + \bar{\delta}_{4i}\eta_4$. The average number of children in the

---

[12] This normalization on the β's and α's is the one used in Pollak and Wales [1980], on which this material is based, and differs from the one used in Chapter 2 ($\sum \alpha_k = 1$). The results are, of course, invariant to the choice of normalization.

[13] Some demand systems may be undefined or imply negative consumption of some goods for certain values of the b's. Fortunately, these problems did not arise in our sample.

[14] For the demand functions to make sense, all of the m's must be positive. Although our linear specification does not constrain the estimates of the m's to be positive, the m's implied by our estimates were positive at all sample points. However, our sample points are group averages, and the m's implied by our estimates are not positive for all possible family compositions. For example, our estimates of the QES imply a negative m for miscellaneous for a family with one child under two and another between two and five. If any of the m's are negative at sample points, then we must either reestimate the system after imposing nonnegativity constraints or use a different functional form for the m's; for example, the form $m_i = \eta^{\varepsilon_i}$ is guaranteed to yield positive values.

household, η_1, does not appear explicitly in this specification because it would be redundant. Since $\eta_1 = \eta_2 + \eta_3 + \eta_4$, we have three rather than four independent demographic variables, and a complete analysis can be based on any three of them. It is convenient to reformulate our specification of demographic effects to include η_1 instead of η_4. To do this, we replace η_4 by $\eta_1 - \eta_2 - \eta_3$ and substitute into the expressions for m_i and b_i. For m_i this yields $m_i = 1 + \varepsilon_{1i}\eta_1 + \varepsilon_{2i}\eta_2 + \varepsilon_{3i}\eta_3, i = 1, 2, 3,$ where $\varepsilon_{1i} = \bar\varepsilon_{4i}$, $\varepsilon_{2i} = \bar\varepsilon_{2i} - \bar\varepsilon_{4i}$, and $\varepsilon_{3i} = \bar\varepsilon_{3i} - \bar\varepsilon_{4i}$. For b_i this yields $b_i = b_i^* + \delta_{1i}\eta_1 + \delta_{2i}\eta_2 + \delta_{3i}\eta_3$ where $\delta_{1i} = \bar\delta_{4i}, \delta_{2i} = \bar\delta_{2i} - \bar\delta_{4i},$ and $\delta_{3i} = \bar\delta_{3i} - \bar\delta_{4i}$.

For what parameter values are consumption patterns independent of household size or composition? In the scaling specification, if the ε's are all equal to zero, then demographic variables do not affect consumption patterns; if all the ε_2's and ε_3's are zero, while the ε_1's are not zero, then consumption patterns depend on household size, but not on household composition. In the translating specification, if all the δ's are zero, then household consumption patterns do not depend on the household's demographic characteristics. If all the δ_2's and δ_3's are zero, while the δ_1's are not zero, then consumption patterns depend on household size but not on household composition.

The number of parameters to be estimated depends on both the underlying functional form of the demand system and on the demographic specification. Take the case of three goods and three independent demographic characteristics. Scaling introduces nine additional parameters to be estimated regardless of the functional form of the demand system. Translating also introduces nine additional parameters to be estimated in the case of the QES and the GTL, demand systems that already include translation parameters. On the other hand in the BTL, a demand system that does not include translation parameters, demographic translating introduces twelve additional parameters.

As in Section 1 we obtain a stochastic specification for these demand systems by adding a disturbance term to the share form of each equation. We denote the 3×1 vector of disturbances corresponding to the ith cell by $u_i = (u_{i1}, u_{i2}, u_{i3})'$ and assume that $E(u_i) = 0$, that $E(u_i u_i') = \Omega$ for all i, and that the u_i are independently normally distributed. Since the dependent variables and the nonstochastic terms in the equations are shares and sum to 1 for each cell, the covariance matrix is singular. Hence, maximum liklihood estimates of the system can be obtained by minimizing the determinant of the sample error covariance matrix with respect to the parameters after dropping any equation.

In Table 2 we present the log likelihood values for the 13 models that we have estimated. We use these results as a basis for discussing three questions: first, do the demographic variables matter? In particular, are consumption patterns significantly affected by the number of children in the household, and, if so, do their ages have a significant additional effect? Second, which of the two methods of incorporating demographic

Table 2   Log Likelihood Values and Regularity Conditions

Model	Method of including variables	Number of parameters	Log likelihood	Regularity conditions
1. QES–no demographic variables	—	8	610.25	69
2. QES–number of children	T	11	696.35	71
3. QES–number of children	S	11	695.36	81
4. QES–number and age of children	T	17	705.69	76
5. QES–number and age of children	S	17	719.93	81
6. BTL–no demographic variables	—	8	597.86	81
7. BTL–number of children	S	11	685.05	81
8. BTL–number and age of children	S	17	706.79	80
9. GTL–no demographic variables	—	11	610.51	81
10. GTL–number of children	T	14	701.19	81
11. GTL–number of children	S	14	703.67	81
12. GTL–number and age of children	T	20	717.85	68
13. GTL–number and age of children	S	20	722.95	81

Notes:
1. T and S indicate whether demographic variables were incorporated through translating or scaling.
2. A common additive constant is omitted from each of the reported log likelihood values.
3. The regularity conditions column displays the number of sample points at which the estimated demand system is well-behaved. The total number of sample points is 81.

variables—translating or scaling—is more consistent with our data? Finally, which of the three demand systems we have estimated—the QES, the BTL, and the GTL—is most consistent with our data?

We consider first the question of whether the number of children and their age distribution affects the consumption patterns of the households in our sample. For all three demand systems, the addition of the number of children to the basic model is significant at the 1% level using both scaling and translating. Similarly the subsequent inclusion of the age distribution variables is significant at the 1% level in all models. Thus it appears that both the number and age of children have a significant effect on consumption patterns, even at the level of our three broad consumption categories, for all three functional forms and for both demographic specifications.

We next compare translating and scaling as alternative methods of incorporating demographic variables into demand systems. From Table 2 we see that scaling results in a higher likelihood value than translating in three of the four comparisons. However, since scaling and translating are not nested specifications, we cannot test the significance of the differences in the likelihood values.

Finally, we compare functional forms. We compare the BTL and the QES functional forms in terms of the values of the likelihood functions reported in Table 2. Since the QES and the BTL (but not the GLT) have the same number of parameters for the case of three goods, a direct comparison of the likelihood functions is of interest; but once again, since the functional forms are not nested, we cannot test the significance of any differences. A comparison of rows 1 and 6, 3 and 7, and 5 and 8 reveals that the QES yields a higher likelihood value than the BTL for all three cases.

We compare the GTL and the BTL using the likelihood ratio test. We find the GTL is a significant improvement over the BTL at the 1% level in all three cases (rows 6 and 9, 7 and 11, and 8 and 13). On the basis of these tests, we have eliminated the BTL from further consideration and have focused on the QES and GTL results.

No meaningful comparison of the QES and the GTL is possible. Since these demand systems are not nested, a formal test cannot be made on the basis of the likelihood values. An informal comparison based on the values of the likelihood function is indecisive: the GTL yields higher likelihood values than the QES, but since it involves three more parameters than the QES, no inference can be drawn from higher likelihood values.

The next set of questions involves the magnitude of the effects of demographic variables on consumption patterns. We first consider the

Table 3 Marginal Budget Shares by Family Size and Expenditure

Expenditure (S. per week)	Consumption category/number of children								
	Food			Clothing			Miscellaneous		
	1	2	3	1	2	3	1	2	3
	QES–Translating								
200	.35	.38	.41	.29	.31	.34	.36	.31	.26
300	.28	.30	.33	.22	.25	.27	.50	.45	.40
400	.20	.23	.25	.16	.18	.20	.64	.59	.55
	QES–Scaling								
200	.37	.40	.41	.29	.30	.31	.34	.31	.28
300	.27	.31	.34	.23	.24	.26	.51	.45	.40
400	.16	.22	.27	.16	.19	.20	.67	.59	.53
	GTL–Translating								
200	.37	.41	.45	.28	.30	.32	.36	.30	.24
300	.27	.31	.34	.22	.24	.26	.50	.45	.39
400	.18	.21	.25	.18	.19	.21	.64	.59	.54
	GTL–Scaling								
200	.38	.43	.48	.28	.30	.31	.33	.27	.21
300	.26	.30	.35	.23	.24	.26	.51	.45	.40
400	.16	.20	.24	.19	.20	.21	.65	.60	.55

Notes:
1. These results correspond to rows 2, 3, 10, and 11 in Table 2.
2. Marginal budget shares are evaluated at 1970 prices. Mean expenditure in 1970 is 315 S. per week.

effect of household size and composition on the marginal budget shares, and then their effects on own- and cross-price elasticities.

In Table 3 we present estimated marginal budget shares corresponding to three levels of family size and three expenditure levels for the GTL and QES systems with scaling and translating; the marginal budget shares are evaluated at 1970 prices.[15] In all four systems, the marginal budget shares for food and clothing increase with family size at all expenditure levels, while that for miscellaneous falls. In addition, the marginal budget shares for food and clothing fall as expenditure rises, while that for miscellaneous rises. Perhaps the most interesting conclusion to be drawn from the table is that the marginal budget share estimates are very similar for both demand specifications and for both methods of incorporating family size.

In Table 4 we report the marginal budget shares corresponding to different family sizes and age distributions for the four models reported in Table 3; the marginal budget shares are all evaluated at 1970 prices, with expenditure of 300 S. per week. For each model, the table reports the marginal budget shares for nine demographic profiles. Specifically, for families with one, two, or three children, we define three age distributions, A_1, A_2, A_3, so that the average age of the children in the family rises in moving from A_1 to A_2 and from A_2 to A_3.[16] In three of the four models, as the average age of the children in the household rises, the marginal budget shares for food and clothing increase (or remain the same), while the marginal budget share of miscellaneous decreases. However, the magnitudes of the marginal budget shares vary with the form of the demand system and the method of incorporating the demographic characteristics. In contrast, when only family size was included, the marginal budget share estimates were not sensitive to the functional form of the demand system or to the method used to incorporate family size.

In Table 5 we report the estimated own-price elasticities corresponding to different family size and age distributions for the four models reported in Tables 3 and 4; the format follows Table 4. As with the marginal budget shares, the own-price elasticities depend on both the form of the demand system and the method of incorporating the demographic characteristics. Each model tells a somewhat different story about own-price elasticities,

---

[15] These comparisons are based on equations that include family size but not the age variables, and hence correspond to rows 2, 3, 10, and 11 in Table 2.

[16] The construction of the A's is as follows. For each family size (η_1 equal to 1, 2, or 3 children) we set the average number of children between the ages of 2 and 5 (η_3) at its sample mean. We then allow the average number of children under 2 (η_2) to take on three values—its sample mean, and approximately its smallest and largest sample values. When η_2 takes on its largest value we denote the age distribution as A_1; when η_2 takes on its mean value we denote the age distribution as A_2; and when η_2 takes on its smallest value we denote the age distribution as A_3. Since this is done for each family size (η_1 equal to 1, 2, or 3 children) the values of η_2 and η_3 corresponding to the A_1, A_2, and A_3 distributions differ for each family size. The range of sample values for the ages of children and their means for each family size are reported in Appendix A.

Table 4 Marginal Budget Shares by Family Size and Age Distribution

| Number of children | Consumption category/age distribution | | | | | | | | |
| | Food | | | Clothing | | | Miscellaneous | | |
	A_1	A_2	A_3	A_1	A_2	A_3	A_1	A_2	A_3
	QES–Translating								
1	.20	.23	.27	.19	.22	.25	.61	.54	.48
2	.24	.27	.29	.22	.25	.28	.54	.48	.42
3	.29	.32	.35	.28	.30	.33	.44	.38	.32
	QES–Scaling								
1	.15	.20	.25	.26	.25	.23	.58	.55	.52
2	.23	.26	.29	.27	.25	.23	.50	.49	.49
3	.29	.31	.31	.28	.26	.24	.43	.43	.45
	GTL–Translating								
1	.11	.20	.30	.18	.22	.27	.71	.58	.42
2	.16	.24	.34	.20	.24	.28	.64	.52	.38
3	.29	.39	.41	.26	.30	.31	.45	.31	.28
	GTL–Scaling								
1	.19	.21	.24	.23	.23	.23	.58	.55	.53
2	.18	.22	.26	.24	.25	.25	.58	.53	.50
3	.32	.35	.38	.27	.27	.27	.41	.38	.35

Note:
1. Marginal budget shares are evaluated at 1970 prices with expenditure of 300 S. per week. Marginal budget shares may not add to unity due to rounding. These results correspond to rows 4, 5, 12, and 13 in Table 2.

although one consistent pattern is the increase (in absolute value) in the price elasticity for food as age increases. This occurs for both models and both methods of incorporating the demographics. Also, in almost all cases the elasticities increase as the number of children increases for both models and for both methods.

Finally we consider the extent to which the 13 models correspond to well-behaved preferences. Table 2 displays the number of sample points at which the Slutsky matrix is negative semidefinite. In 8 of the models regularity conditions hold at all 81 sample points and in all other models they hold at over 80% of the data points. The translog models overall appear to be slightly superior to the QES on these grounds, although if just the models that include age and number of children are considered, the QES is marginally better. Returning to the comparison between scaling and translating, note that in all four cases regularity conditions are satisfied for at least as many sample points with scaling as with translating.

In summary we have studied two topics in empirical demand analysis: the specification of functional forms for the demand equations and two procedures for incorporating demographic variables. We found that both

Table 5   Own-Price Elasticities by Family Size and Age Distribution

Number of children	Consumption category/age distribution								
	Food			Clothing			Miscellaneous		
	A_1	A_2	A_3	A_1	A_2	A_3	A_1	A_2	A_3
				QES–Translating					
1	−.21	−.24	−.37	−.22	−.24	−.54	−.63	−.56	−.73
2	−.25	−.36	−.54	−.25	−.53	−1.04	−.56	−.74	−1.14
3	−.49	−.72	−1.02	−.95	−1.71	−2.80	−1.11	−1.77	−2.64
				QES–Scaling					
1	−.53	−.62	−.69	−1.34	−1.43	−1.54	−.81	−.84	−.88
2	−.65	−.71	−.78	−1.47	−1.58	−1.75	−.88	−.92	−.98
3	−.68	−.72	−.79	−1.55	−1.68	−1.94	−.92	−.98	−1.05
				GTL–Translating					
1	−.81	−1.10	−1.42	−1.07	−1.23	−1.42	−1.00	−1.59	−2.25
2	−.92	−1.17	−1.47	−1.17	−1.32	−1.52	−1.34	−1.94	−2.60
3	−1.24	−1.55	−2.11	−1.46	−1.70	−2.21	−2.48	−3.34	−5.41
				GTL–Scaling					
1	−.76	−.78	−.81	−1.31	−1.23	−1.16	−1.32	−1.34	−1.34
2	−.80	−.84	−.87	−1.36	−1.31	−1.25	−1.56	−1.56	−1.55
3	−.89	−.91	−.93	−1.36	−1.30	−1.25	−1.43	−1.38	−1.31

Note: See Table 4.

the number and age of children in a family have a significant effect on consumption patterns, regardless of whether the underlying demand system is assumed to be QES or translog, and regardless of whether the demographic effects are assumed to operate through translating or through scaling.

Although demographic translating and demographic scaling are not nested specifications, some comparison of these two procedures is possible. In three of the four comparisons, scaling resulted in a higher likelihood value; and in all four cases, scaling did at least as well as translating in terms of regularity conditions.

We found the GTL significantly superior to both the linear translog and the BTL. This suggests that the translog forms previously discussed in the literature are more restrictive than has generally been recognized, and that more flexible responses to changes in expenditure are required, at least for that analysis of household budget data.

The QES and the translog specifications can be compared using the values of the likelihood functions, provided the specifications have the same number of parameters. Using this likelihood function criterion, the QES was superior to the translog in all of the cases we can compare. The highest likelihood values were attained by the GTL, but since this demand system contains three more parameters than the QES, we were unable to compare these systems.

3. DEMOGRAPHIC SPECIFICATIONS

In this section we illustrate the use of the five general procedures discussed in Chapter 3 for incorporating demographic variables in demand analysis. Due to the complexity of some of these procedures we estimate the generalized CES demand system rather than the QES or BTL. To simplify the analysis further we consider just one demographic variable, the number of children in the household. We assume that this variable enters the functions specifying the five procedures in a linear manner. The sample consists of data for the period 1966 through 1972 and yields 108 cells, which, as before, we treat as consumption patterns of households rather than as cell means.

The generalized CES demand equations (in share form) are given by

$$(6) \qquad w_i = \frac{p_i b_i}{\mu} + \frac{\alpha_i^c p_i^{1-c}}{\sum \alpha_k^c p_k^{1-c}} [1 - (\sum p_k b_k)/\mu], \qquad \sum \alpha_k = 1,$$

where the α's, b's, and c are parameters to be estimated. The parameter c is the elasticity of substitution between "supernumerary quantities," $(x_i - b_i)$. We use the same three consumption categories as in Sections 1 and 2; with three goods the generalized CES contains six independent parameters.

As usual we obtain a stochastic specification for this demand system by adding a disturbance term to the share form of each demand equation. We denote the 3×1 vector of disturbances corresponding to the ith cell by $u_i = (u_{i1}, u_{i2}, u_{i3})$ and assume $E(u_i) = 0$, $E(u_i u_i') = \Omega$ for all i, and that the u_i are independently normally distributed. Once again this is our standard stochastic formulation with log likelihood given by (3) in Chapter 5.

Table 6 presents log likelihood values and some other statistics for seven specifications: the five procedures described in Chapter 3 for which demand equations are given in Appendix B; the "pooled" specification that combines data from different family types and estimates a single demand system, implicitly assuming that consumption patterns are independent of demographic variables; and the "unpooled" specification that estimates three separate demand systems, one for each family type, implicitly assuming that demographic variables affect all demand system parameters.[17]

From Table 6 it is clear that family size significantly affects consumption patterns regardless of the method used to incorporate it. Comparing the

---

[17] In the "unpooled specification" the family types are households with one child, two children, and three or more children. Since data are available on households with exactly three children in one year only (1972), these data are used together with those for households with four or more children in 1972 to form the family type consisting of three or more children.

Table 6 Log Likelihood Values

	Translating	Scaling	Gorman	Reverse Gorman	Modified Prais–Houthakker	Unpooled	Pooled
Log likelihood	861.6262	863.4707	864.4165	864.5088	864.5157	870.5274	772.212
Estimate of v	1	0	1.35	1.32	n/a	n/a	n/a
Estimate of c	2.57	2.31	2.66	2.63	2.58	1.66, 4.00, 2.09	2.88
Number of estimated parameters	9	9	10	10	9	18	6
Chi-square	17.80	14.11	12.22	12.04	12.02	n/a	196.63

Notes:
1. n/a indicates not applicable.
2. For translating and scaling the value of v is not estimated but is set to 1 and 0, respectively.
3. For the unpooled estimates, the values for c correspond to families with 1 child, 2 children, and 3 or more children.
4. The chi-square value is calculated as minus twice the difference between 870.5274 and the log likelihood value in the column.

pooled results with those obtained using each of the five procedures, we find a very substantial decrease in the value of the likelihood in each case. Further, comparing each of the five procedures with the unpooled specification, the decrease in the likelihood function is significant at the 5% level only for demographic translating.[18] Because these results confirm the validity of incorporating demographic variables into the generalized CES by means of our five procedures we do not report marginal budget shares or price elasticities for the pooled or the unpooled specifications.

All five procedures yield estimated parameters that correspond to well-behaved preferences in all or virtually all of the 108 price–expenditure–demographic situations represented in our data.[19] Further, all yield similar values of c, the elasticity of substitution between supernumerary quantities: estimates range from 2.31 to 2.66.[20, 21]

The modified Prais–Houthakker procedure makes a very strong showing against the other four procedures on the basis of the value of the likelihood function: the 9-parameter modified Prais–Houthakker procedure yields a value of the likelihood function that is greater than that of the other two 9-parameter procedures (translating and scaling), as well as that of the two 10-parameter procedures (Gorman and reverse Gorman).[22] No formal classical tests are possible, however, because none

---

[18] The decrease is not significant for demographic translating at the 1% level.

[19] For translating, scaling, and the Gorman procedures, the Slutsky matrix was negative semidefinite in all situations; for the reverse Gorman and modified Prais–Houthakker procedures, it was negative semidefinite in 107 of 108 situations.

[20] As Table 6 shows, the unpooled estimates exhibit a wider range.

[21] For all five procedures, the generalized CES is significantly superior to the familiar LES at the 5% level. The generalized CES reduces to the LES when c = 1, and we have tested whether our estimated value of c differs significantly from 1. With the LES, the modified Prais–Houthakker procedure yields a likelihood value of 862.5833, while the other four procedures imply identical estimating equations and yield a likelihood value of 859.6875. It is easy to verify that scaling and translating are equivalent for the LES. Their equivalence to the Gorman and reverse Gorman procedures is less obvious but follows from the observation that v is not identified when c = 1. To see this, consider Eq. (B3) in Appendix B: when c = 1, the marginal budget shares are constant and the α's and v enter only in terms of the form $[1 + (1 - v)\alpha_i \eta]b_i + v\alpha_i \eta$. Rewriting these as $b_i + \alpha_i[v + b_i[v + b_i(1 - v)]]\eta$ and denoting $\alpha_i[v + b_i(1 - v)]$ by π_i, it is clear that estimates of b_i and π_i, but not v, can be obtained: all values of v yield the same likelihood values, since as v changes, the α_i's can adjust to give the same estimates of π_i.

[22] We also estimated the special case of the modified Prais–Houthakker procedure in which the specific scales are the same for all goods, and the even more restrictive special case in which their common value is such that children receive the same weight as adults. Since our basic units of observation are households with two adults and at least one child, this implies a coefficient of 0.5, so that a household with four children (i.e., six persons) spending 400 S. per week purchases twice as much of every good as a household with one child (i.e., three persons) spending 200 S. per week. The likelihood values for these models are 860.5457 and 814.2324, respectively; at the 5% level, these restrictions are rejected against the unpooled specification.

of the other four procedures is nested in the modified Prais–Houthakker procedure, nor is it nested in any of them.[23]

Scaling compares favorably with the remaining three procedures. It has a higher likelihood value than translating, which is also a 9-parameter procedure. It is not rejected at the 5% level against the unpooled specification nor against either the Gorman or reverse Gorman procedures.[24]

The reverse Gorman procedure is superior to the Gorman on the basis of the values of their likelihood functions. Since both procedures involve ten parameters, this provides a plausible basis for comparison, although no formal classical tests are possible.[25]

Translating is the weakest of the five procedures. It has a lower likelihood value than the modified Prais–Houthakker procedure and scaling, which also have nine parameters, and it is rejected at the 5% (but not at the 1%) level against the Gorman and reverse Gorman procedures.[26]

Table 7 presents marginal budget shares and own-price elasticities implied by our estimates of each of the five procedures for households with one, two, and three children. Price elasticities and marginal budget shares are evaluated at 1970 prices and an expenditure level that is close to the median expenditure for households in 1970.

The marginal budget share estimates implied by the five procedures are strikingly similar. The four procedures other than translating permit marginal budget shares to vary with family size; all show that the marginal budget share for food increases slightly with family size, that for clothing is almost constant, while that for miscellaneous decreases slightly. In the generalized CES, translating necessarily yields marginal budget shares that are independent of the number of children. The marginal budget

---

[23] Although techniques have been proposed for testing and comparing nonnested hypotheses (e.g., Pesaran and Deaton [1978]; Davidson and MacKinnon [1981]; Pollak and Wales [1991]), their application is beyond the scope of this book.

[24] Muellbauer [1977] rejects scaling against the unpooled specification as a method for incorporating household size into demand analysis. In addition to using a different functional form, Muellbauer uses data for a somewhat different period (1968–1973), includes households with no children, and uses ten consumption categories, some of which include consumer durables.

[25] Both the Gorman and reverse Gorman procedures yielded estimates of v outside the range [0, 1]. This does not violate regularity conditions, nor does it imply predicted responses to changes in the demographic variables very different from those implied by the other procedures. The likelihood functions that correspond to the Gorman and reverse Gorman procedures each had two maxima, a local maximum corresponding to a value of v below 0, and a global maximum corresponding to a value greater than 1. The values corresponding to the global maxima are given in Table 6. The estimated value of v at the local maximum for the Gorman (reverse Gorman) procedure is $-.60$ ($-.13$) and the corresponding value of the likelihood function is 864.3657 (864.1946).

[26] In Section 2 we found translating to be generally inferior to scaling for the QES and GTL, both of which are nonlinear in expenditure. Thus, it seems unlikely that translating performs poorly in Table 6 merely because the generalized CES is linear in expenditure.

Table 7 Marginal Budget Shares and Own-Price Elasticities by Family Size:
Alternative Procedures for Incorporating Demographic Effects

| | Consumption category/number of children | | | | | | | | |
| | Food | | | Clothing | | | Miscellaneous | | |
Procedure	1	2	3	1	2	3	1	2	3
	Marginal budget shares								
Translating	.29	.29	.29	.23	.23	.23	.48	.48	.48
Scaling	.27	.29	.30	.23	.23	.23	.50	.48	.47
Gorman	.27	.29	.31	.23	.22	.23	.50	.49	.47
Reverse Gorman	.27	.29	.30	.23	.22	.23	.50	.49	.47
Modified Prais– Houthakker	.27	.29	.31	.22	.22	.23	.51	.49	.46
	Own-price elasticities								
Translating	−.78	−.75	−.72	−1.43	−1.49	−1.56	−1.41	−1.51	−1.65
Scaling	−.81	−.76	−.71	−1.64	−1.55	−1.44	−1.57	−1.56	−1.54
Gorman	−.73	−.75	−.76	−1.43	−1.50	−1.58	−1.42	−1.52	−1.66
Reverse Gorman	−.73	−.74	−.76	−1.44	−1.50	−1.57	−1.42	−1.52	−1.65
Modified Prais– Houthakker	−.78	−.75	−.72	−1.44	−1.51	−1.46	−1.53	−1.53	−1.53

Note:
1. Price elasticities are evaluated at 1970 prices (all equal to 100) and an expenditure level of 300 S. per
    week. Marginal budget shares are also evaluated at 1970 prices, but are independent of expenditure.

share estimates for translating are essentially identical to those implied
by the other four procedures for two-child households.

A comparison of estimated own-price elasticities also suggests that the
differences among procedures are small, especially for households with
two children. For one-child households, the largest difference among
procedures occurs for clothing, but this difference is only .21; for
three-child families, the largest difference, also for clothing, is .14. We have
calculated both own- and cross-price elasticities corresponding to
expenditure levels of 200, 300, and 400 S. per week at 1970 prices, although
we report only own-price elasticities at 300 S. per week. Estimated
cross-price elasticities are very similar for all five procedures and the size
of these differences decreases with expenditure: the largest difference
between comparable elasticities at 300 S. per week is only .15.

In summary, we have estimated and compared five procedures for
incorporating demographic variables into complete demand systems. Our
comparisons are based on a generalized CES demand system into which
we incorporated a single demographic variable, the number of children
in the household. We have also estimated the pooled specification in which
demographic variables have no effect on consumption patterns, and the
unpooled specification in which they affect all demand system parameters.

We rejected the pooled specification against each of the five procedures,
indicating that the number of children does affect consumption patterns

independently of the procedure used. Of the five procedures only demographic translating could be rejected against the unpooled specification, indicating that the other four procedures are reasonably consistent with the data. These four procedures imply similar responses to changes in prices, total expenditure, and the number of children. Although no formal ranking of these procedures is possible, statistical tests can be applied to pairs of procedures that are nested, and the likelihood value provides a plausible basis for comparing procedures with the same number of independent parameters. Using these criteria, translating made the weakest showing while the two Gorman procedures were dominated by scaling, and the modified Prais–Houthakker was best of all.

4. DYNAMIC AND STOCHASTIC STRUCTURE

In this section we augment the QES model by introducing dynamic and stochastic structure. Our dynamic structure allows for the possibility of taste change by assuming that some utility function parameters are time-dependent. Our stochastic structure allows disturbances across households to be correlated in a given year. Unlike the stochastic specification used in preceding sections, the error components structure we use here is consistent with the possibility that, for example, in a particularly cold year most households consume more fuel than usual. We consider first the dynamic and then the stochastic specification.

We estimate four dynamic specifications: (a) a linear time trend specification in which some demand system parameters (the b's) vary linearly with time; (b) a lagged consumption specification in which some demand system parameters (again the b's) vary linearly with a variable representing past consumption; (c) a constant tastes specification in which all demand system parameters remain fixed; and (d) a model in which both (a) and (b) hold. The time trend specification is one of exogenous taste change and can be written as

$$(7) \qquad b_{it} = b_i^* + \beta_i t.$$

The lagged consumption specification is one of endogenous taste change and can be written as

$$(8) \qquad b_{it} = b_i^* + \beta_i z_{it-1},$$

where z_{it-1} is a variable representing past consumption. As discussed in Chapter 4 the interpretation of the lagged consumption specification depends on the variable chosen to represent past consumption. For example, if z_{it-1} represents the household's own consumption of the ith good in period $t-1$, then the implied model is one of habit formation; if z_{it-1} represents other households' consumption of the ith good in period $t-1$, then the specification is one of interdependent preferences. Incorporating both the time trend and lagged consumption into the model yields

$$(9) \qquad\qquad b_{it} = b_i^* + \beta_i t + \gamma_i z_{it-1}.$$

All of these models contain the constant tastes specification as a special case in which the β's (and/or γ's) are set to 0.

The usual "independent" stochastic specification adds disturbance terms to the share equations and assumes these disturbances are independent across households and over time. In addition to this independent specification we consider an error components structure in which the disturbance term is the sum of two components, one independent across households and over time, and the other a "time-specific effect" (TSE), which is the same for all households in a particular year.[27] The independent specification is thus a special case of the error components structure in which the TSE is absent. The TSE allows a positive correlation between the disturbances of different households in a particular year.

To formalize the error components model, we let u_{rt} denote the $(n-1) \times 1$ vector of disturbances added to $n-1$ of the equations in the demand system (1), where r denotes the household and t the time period. These disturbances are the sum of two components

$$(10) \qquad\qquad u_{rt} = e_t + \varepsilon_{rt}, \qquad \begin{matrix} r = 1, \dots, q_t \\ t = 1, \dots, T, \end{matrix}$$

where q_t is the number of households in period t and T the number of time periods. We do not assume that household r in period t is the same as household r in period τ nor even that there are the same number of households in every period. We do assume that the $(n-1) \times 1$ disturbance vector e_t and ε_{rt} are independently normally distributed with zero means and covariance matrices given by Γ and Δ, respectively, where Γ is positive semidefinite and Δ positive definite. The covariance matrix for the u's is thus given by

$$(11) \qquad\qquad E(u_{rt}u_{s\tau}) = \begin{cases} \Gamma + \Delta, & r = s, & t = \tau \\ \Gamma & r \neq s, & t = \tau \\ 0 & & t \neq \tau. \end{cases}$$

Under these assumptions the log of the likelihood is (aside from an additive constant)

$$(12) \qquad L = -\frac{1}{2}\sum_{t=1}^{T} \left\{ \log |W_t| + (q_t - 1)\log|\Delta| + \sum_{s=1}^{q_t} u'_{st}\Delta^{-1}u_{st} \right.$$
$$\left. + \sum_{s=1}^{q_t}\sum_{r=1}^{q_t} u'_{st}\frac{(W_t^{-1} - \Delta^{-1})}{q_t}u_{rt} \right\},$$

---

[27] Since we do not have data on the same households in successive years, we do not discuss three-component structures involving household-specific disturbance terms.

where $W_t = \Delta + q_t \Gamma$.[28] Since Γ and Δ are $n - 1 \times n - 1$ symmetric matrices this stochastic specification contains $n(n - 1)$ independent parameters.[29]

We estimate the same QES form as in Sections 1 and 2, using data for 1967–1972.[30] As mentioned earlier we estimate separate models for families with one, two, and three or more children. To represent past consumption we use the average consumption of all households in the sample in the previous year. Because cell sizes vary, it is a weighted average of cell means

$$(13) \qquad\qquad z_{it} = \frac{\sum_{r=1}^{q_t} \eta_{rt} x_{irt}}{\sum_{r=1}^{q_t} \eta_{rt}},$$

where x_{irt} is consumption of the ith good by income group r in period t and η_{rt} denotes the number of households in income group r in period t.[31]

Table 8 contains log likelihood values (aside from an additive constant) for the various models. Perhaps the most interesting result is the statistical insignficance of the TSE in all cases. The TSE involves an additional three parameters, while the 5% critical value for the chi-square distribution with three degrees of freedom is 7.81. In no case does twice the difference in the log likelihoods exceed this value. Indeed in six of the comparisons the TSE results in no increase in the likelihood function; in these cases the constraint that Γ be positive definite results in all three of the TSE parameters being set to zero in the likelihood maximization. We consider next the dynamic aspects of the model.

The significance of the dynamic effects depends on the number of children in the household. For households with one child the static model

---

[28] The derivation of this likelihood function appears in Chapter 5, while a more extensive analysis of this type of error components model appears in Magnus [1982].

[29] We impose the restriction that Δ be positive definite by writing it as the product of a lower triangular matrix (ΔL) and its transpose, and estimate the elements of ΔL rather than Δ. The same procedure is followed for Γ, thus giving elements of ΓL.

[30] One year is dropped to allow for lagged consumption.

[31] We ignore aggregation problems that may call for higher moments of lagged consumption. In a habit formation specification, the appropriate variable is the average lagged consumption of the households in the cell; in an interdependent preferences specification, the appropriate variable is the average lagged consumption of households assumed to influence those in the cell. The only data we have are those reported in the previous year's sample, and because the samples are not panels, these are different households. Using average consumption in the previous year's sample is compatible with an interdependent preferences specification in which tastes depend on the average past consumption of all households with the same demographic profile, regardless of their income or expenditure. The connection with a habit formation specification is rather loose.

An alternative but substantially more complicated approach would relate the preferences of households in each percentile of the income or expenditure distribution to the consumption of those occupying that position in the previous year.

Table 8   Log Likelihood Values for Dynamic QES Specifications

Number of children	Time-specific effect	Static	Time trend	Lagged consumption	Lagged consumption and time trend
1	No	241.97	242.04	242.31	243.42
	Yes	241.97*	242.04*	242.31*	243.42*
2	No	253.41	257.04	254.48	260.82
	Yes	254.58	259.49	256.48	260.83
3 or more	No	230.50	237.76	236.65	241.91
	Yes	232.61	238.23	236.65*	241.91*

Notes:
1. The asterisk indicates cases in which including the TSE resulted in no increase in the likelihood value, as discussed in the text.

is acceptable, with neither the time trend nor lagged consumption having a significant impact on consumption patterns. For households with two children neither the time trend nor lagged consumption is significant individually, but they are marginally significant jointly. That is, starting with the full model we reject the hypothesis of constant tastes at the 5% level. Households with three or more children exhibit the most significant dynamic behavior, with both the time trend and lagged consumption individually and jointly highly significant. It is not clear to us why the dynamic element becomes increasingly significant with family size.

The habit formation or interdependent preferences interpretation of the lagged consumption model requires positive β's. For households with three or more children all three β's are negative in the full model (and when the time trend is omitted), while for two-children households only the β for food is negative. It is tempting to interpret negative β's as reflecting partial stock adjustment for consumer durables rather than habit formation (see Houthakker and Haldi [1960]), although the absence of interest rates suggests that the interpretation is rather loose. The consumer durable interpretation would be more convincing if the β''s for clothing were negative for all family sizes.

We present marginal budget shares and own-price elasticities in Table 9 for our preferred models. For one-child families we base our results on the static model, while for the other families we base them on the full model containing both time-trend and lagged-consumption terms. The results are presented for 1970 with expenditure levels given by the sample points for that year.

Marginal budget shares follow the same pattern as a function of expenditure for all family sizes. The marginal budget share for food falls sharply, and that for miscellaneous rises sharply, as expenditure rises. The share of clothing falls sharply for all families except those with two children,

Table 9 Marginal Budget Shares and Own-Price Elasticities: Selected QES Specifications

Expenditure (S. per week)	Marginal budget shares/ consumption category			Own-price elasticities/ consumption category		
	Food	Clothing	Miscel-laneous	Food	Clothing	Miscel-laneous
	Families with one child; static model					
220.0	.35	.26	.39	−.59	−1.08	−.72
242.2	.33	.25	.42	−.61	−1.14	−.79
284.8	.28	.23	.49	−.65	−1.26	−.88
399.5	.14	.17	.69	−.77	−1.61	−1.03
	Families with two children; time trend and lagged consumption					
230.7	.40	.23	.37	−.74	−1.28	−1.00
259.0	.37	.23	.40	−.82	−1.46	−1.13
280.2	.35	.23	.42	−.87	−1.59	−1.22
320.2	.31	.23	.46	−.99	−1.83	−1.37
359.3	.27	.22	.51	−1.11	−2.07	−1.50
476.2	.16	.21	.63	−1.55	−2.78	−1.79
	Families with three or more children; time trend and lagged consumption					
240.1	.42	.33	.25	−1.05	−3.09	−2.72
291.2	.38	.29	.33	−.58	−.94	−1.02
331.9	.34	.26	.40	−.37	−.37	−.52
480.6	.19	.15	.66	−.48	−.79	−1.17

for which it falls only slightly as income rises. These results are consistent with those in earlier sections.

The pattern of own-price elasticities as a function of expenditure depends on family size. These elasticities increase (in absolute value) as income rises for families with one and two children, but decrease for families with three or more children. The pattern for one- and two-child families is consistent with the results reported in Section 1 based on estimates that did not take family size into account.

Our estimated utility functions satisfy the regularity conditions of economic theory in almost all cases. For our preferred models the Slutsky matrix is negative semidefinite at all sample points for families with one or two children (a total of 64), and at 29 of 32 sample points for families with three or more children.

In summary we have explored two extensions to the models discussed in previous sections. First, we have allowed for the possibility of taste change, and find this significant for families with two or more children. Second, we have allowed disturbances across households to be correlated in a given year; but in no cases are these error components effects significant.

5. RANDOM COEFFICIENTS MODELS

In this section we consider random coefficients models for some of our functional forms. In Chapter 5 we showed that for the QES and LES a random coefficients stochastic structure implies share equation disturbances with a multivariate normal distribution; for these functional forms, such a stochastic structure is computationally tractable. For the BTL and GTL, on the other hand, the share equation disturbances resulting from a random coefficients specification involve ratios of normal variables; estimation of such models is beyond the scope of this book. In this section we use the same data set as in Section 2 to estimate a random coefficients model for the QES, with age and number of children incorporated either through scaling or translating. We also present results for the LES, although the restrictions imposed by the LES are clearly rejected by the data.

The QES demand equations in share form are given by (5) above. For reasons discussed in Chapter 5 we assume that the c's are randomly distributed across the sample. Hence the c_i's are replaced in (5) by $c_i + \varepsilon_i$ to give

$$(14) \quad w_i = \frac{p_i b_i}{\mu} + a_i \left(1 - \sum \frac{p_k b_k}{\mu} \right)$$

$$+ \left\{ \frac{p_i}{\mu}(c_i + \varepsilon_i) - a_i \sum \frac{p_k}{\mu}(c_k + \varepsilon_k) \right\} \prod \left(\frac{p_k}{\mu} \right)^{-2a_k} \left(1 - \sum \frac{p_k b_k}{\mu} \right)^2.$$

We assume that the covariance matrix for the ε's satisfies

$$(15) \qquad E(\varepsilon_i \varepsilon_j) = \frac{\sigma_{ij} \mu^{2\alpha}}{p_i^\alpha p_j^\alpha \delta^2(P, \mu)}, \qquad i = j$$

$$= 0, \qquad i \neq j,$$

where $\delta(P, \mu) \equiv \prod \left(\frac{p_k}{\mu} \right)^{-2a_k} \left[1 - \sum \frac{p_k b_k}{\mu} \right]^2$ and α is a parameter to be estimated. Further the ε's are assumed to be normally distributed and uncorrelated across observations, and the σ_{ij} are constants. Denoting the disturbance for the ith share equation at observation t in (14) as v_{it} and the corresponding disturbance covariance matrix as Ω_t, the ijth element of Ω_t is given by

$$(16) \quad E(v_{it} v_{jt}) = \frac{p_i p_j \sigma_{ij} - p_i a_j \sum p_k \sigma_{ki} - a_i p_j \sum p_k \sigma_{kj} + a_i a_j \sum \sum p_k p_\ell \sigma_{k\ell}}{p_i p_j},$$

where for simplicity the t subscripts have been dropped from the p's.

The LES demand equations in share form are given by (14) above except that the last term on the right-hand side is set to zero. We assume that the b's are randomly distributed and replace the b_i by $b_i + \varepsilon_i$ to give

(17) $$w_i = \frac{p_i}{\mu}(b_i + \varepsilon_i) + a_i\{1 - \sum \frac{p_k}{\mu}(b_k + \varepsilon_k)\}.$$

If we assume

(18) $$E(\varepsilon_i \varepsilon_j) = \frac{\sigma_{ij}\mu^{2\alpha}}{p_i^\alpha p_j^\alpha}, \qquad i = j$$
$$= 0, \qquad i \neq j$$

and retain the other assumptions about the ε's that we made for the QES, then the disturbance covariance matrix for (17) will again be Ω_t as defined in (16). These covariance matrices are the same because, as we noted in Chapter 5, the b's enter the LES in essentially the same way that the c's enter the QES.

The log likelihood function for a sample of T observations (aside from an additive constant) is then given by (40) in Chapter 5 for both the LES and QES. For simplicity we assume that $\sigma_{ij} = 0$ for $i \neq j$. Since Ω depends on t, the likelihood function cannot be concentrated with respect to these σ_{ii} parameters. As discussed in Chapter 5 this introduces n covariance parameters (σ_{ii}) to be estimated; with $n = 3$ this is the same number of covariance parameters as appears in the standard additive model.[32]

We incorporate the age and number of children into the nonstochastic part of (14) through the use of demographic translating and demographic scaling as in Section 2.

In Table 10 we present the log likelihood values and number of sample points (out of 81) at which the estimated demand systems satisfy the regularity conditions. A number of interesting conclusions can be drawn. First, using standard likelihood ratio tests in all cases we find the value of α significantly different from zero, but not significantly different from one. This corresponds with our a priori beliefs, as discussed in Chapter 5, that the disturbances associated with the share equations will be homoskedastic as a function of income, whereas those associated with the expenditure equations will not be. In view of this finding we consider below only the random coefficients model based on $\alpha = 1$. A comparison of this model with the standard additive disturbance model reveals likelihood values that are virtually identical in all three cases, as indeed are the number of sample points satisfying regularity conditions. Further, although not reported here, the simultaneous inclusion of an additive disturbance and random coefficients disturbance never results in a significantly higher likelihood value than is obtained with the inclusion of

---

[32] The standard additive model includes $n(n-1)/2$ covariance parameters. We have also estimated the more general random coefficients models in which $\sigma_{ij} \neq 0$, $i \neq j$, but we do not report the results below since in no case are these additional parameters sgnificant.

Table 10 Log Likelihood Values and Regularity Conditions

	Log likelihood	Regularity conditions
	QES	
Demographic scaling		
Additive disturbance model	710.93	81
Random coefficients model		
$\alpha = 1$	710.25	81
$\alpha = 0$	707.20	81
α estimated (.64)	711.51	81
Demographic translating		
Additive disturbance model	705.69	76
Random coefficients model		
$\alpha = 1$	705.56	77
$\alpha = 0$	697.61	77
α estimated (.85)	705.81	78
	LES	
Additive disturbance model	671.66	81
Random coefficients model		
$\alpha = 1$	671.25	81
$\alpha = 0$	663.89	81
α estimated (1.003)	671.25	81

Notes:
1. The regularity conditions column records the number of sample points (out of 81) at which estimated demand systems satisfy the regularity conditions.
2. For the LES, translating and scaling are equivalent procedures.

just one type of disturbance. This suggests that there is little to choose between the two methods of incorporating the random component, and no need to use both.

In Table 11 we report marginal budget shares by family size and age composition for the additive disturbance and random coefficients models. The additive disturbance results for the QES are the same as those reported in Table 4 and are repeated here for ease of comparison with the random coefficients model. As is clear from the table, the two methods yield identical estimates in all cases (to two decimal places). Similar results hold for the own-price elasticities and are not presented here, since for the random coefficients model they are virtually identical to those given in Table 5 for the additive disturbance model.

In summary we have estimated random coefficients models for the LES and QES. These models are computationally tractable if the b's in the LES or the c's in the QES are assumed to be stochastic. Reasonable assumptions about the behavior of the share equation error variances as functions of income and prices yield the same disturbance covariance

Table 11 Marginal Budget Shares by Family Size and Age Distribution: Additive Disturbance and Random Coefficients Models

| Number of children | Consumption category/age distribution | | | | | | | | |
| | Food | | | Clothing | | | Miscellaneous | | |
	A_1	A_2	A_3	A_1	A_2	A_3	A_1	A_2	A_3
	QES–Scaling—Additive Disturbance Model								
1	.15	.20	.25	.26	.25	.23	.58	.55	.52
2	.23	.26	.29	.27	.25	.23	.50	.49	.49
3	.29	.31	.31	.28	.26	.24	.43	.43	.45
	QES–Scaling—Random Coefficients Model								
1	.15	.20	.25	.26	.25	.23	.58	.55	.52
2	.23	.26	.29	.27	.25	.23	.50	.49	.49
3	.29	.31	.31	.28	.26	.24	.43	.43	.45
	QES–Translating—Additive Disturbance Model								
1	.20	.23	.27	.19	.22	.25	.61	.54	.48
2	.24	.27	.29	.22	.25	.28	.54	.48	.42
3	.29	.32	.35	.28	.30	.33	.44	.38	.32
	QES–Translating—Random Coefficients Model								
1	.20	.23	.27	.19	.22	.25	.61	.54	.48
2	.24	.27	.29	.22	.25	.28	.54	.48	.42
3	.29	.32	.35	.28	.30	.33	.44	.38	.32

Notes:
1. See note to Table 4.
2. For the LES the marginal budget shares are constant and are .25, .24, and .51 for food, clothing, and miscellaneous for both the additive error model and the random coefficients model.

matrix in both of these cases. There are two major findings: first, the share but not the expenditure equation disturbances are homoskedastic with respect to income; and second, the random coefficients model yields virtually the same results as does the additive disturbance model.

APPENDIX A: DATA

As mentioned in the text, we analyze three broad consumption categories in this chapter: food, clothing, and miscellaneous. The categories are defined as follows. The food category does not include alcoholic drink. Clothing is officially "clothing and footwear." Miscellaneous is the sum of two categories from the survey, "other goods" and "services." The principal subcategories of other goods are "leather, travel and sports goods; jewelry; fancy goods, etc.," "books, magazines and periodicals," "toys and stationery goods, etc.," and "matches, soap, cleaning materials, etc." The principal subcategories of services are "radio and television, licenses and rentals," "educational and training expenses," and "subscriptions and donations;

hotel and holiday expenses; miscellaneous other services." The survey reports seven major expenditure categories that we have omitted entirely: "housing," "fuel, light, and power," "alcoholic drink,"tobacco," "durable household goods," "transport and vehicles," and "miscellaneous." Our three categories of food, clothing, and miscellaneous account for between 46 and 58% of total consumption expenditures.

Mean expenditures on the three consumption categories are available by income level and number of children (on a consistent basis) for the period 1966 through 1972. This yields 108 cells, which we used as basic data points in the equation estimations. In Section 1 only the two years 1966 and 1977 were used; in Sections 3 and 4, the seven years 1966–1972 were used. In Sections 2 and 5, the five years 1968–1972 were used since information on the age distribution of children was not available for 1966 and 1967. A tabulation of the number of cells by year, number of children, and number of income levels is given below.

Classification of Basic Data Cells

Year	Number of children in household	Number of income levels	Total number of cells
1966	1, 2, 3 or more	4	12
1967	1, 2, 3 or more	5	15
1968	1, 2, 3 or more	5	15
1969	1, 2	5	10
	3 or more	4	4
1970	1, 3 or more	4	8
	2	6	6
1971	1, 2, 3 or more	6	18
1972	1, 2	6	12
	3	5	5
	4 or more	3	3
			108

The range and sample mean for the age variables by number of children in the household are as follows (for years 1968–1972):

Number of children in household	Age under 2	Age 2–4	Age 5–16
1	.10–.51 (.33)	.11–.25 (.20)	.31– .65 (.47)
2	.11–.55 (.30)	.32–.83 (.60)	.62–1.57 (1.10)
3	.10–.50 (.29)	.47–.81 (.63)	.69–2.41 (2.08)

For the results in Tables 4 and 5 we have set the "2–4" variable (η_3) at its mean of .20 and the "under 2" variable (η_2) at .16, .33, and .50, for a one-child household. For a two-child household the corresponding values are .60 for η_3 and .15, .30, and .45 for η_2; for a three-child household the values are .63 for η_3 and .15, .29, and .45 for η_2.

All households consist of one man, one woman, and at least one child.

The price indexes for food and clothing are taken directly from Table 4.9 of the 1974 *Annual Abstract of Statistics* (U.K. Central Statistical Office). The price index for the miscellaneous category is a weighted average of the "miscellaneous goods" and "services" indexes in the table, with the weights given by the table entries for 1966. Price indexes for all three categories are set to unity in 1970 and are given below.

Price Indexes

Year	Food	Clothing	Miscellaneous
1966	.825	.888	.787
1967	.846	.902	.811
1968	.879	.916	.866
1969	.935	.951	.926
1970	1.000	1.000	1.000
1971	1.111	1.068	1.109
1972	1.209	1.145	1.170

In Pollak and Wales [1978], we used price indexes obtained from Tables 29 and 30 of *National Income and Expenditure 1964–74* (U.K. Central Statistical Office, 1975). We were unaware of the existence of the retail price series at that time. For this reason and because we are now using the Σ-QES, the results reported in Section 1 differ from those reported in Pollak and Wales [1978].

APPENDIX B: GENERALIZED CES DEMAND SYSTEMS WITH DEMOGRAPHIC VARIABLES

Introduction of the number of children in the household (η) into the generalized CES demand system (6) using the five methods discussed in Section 3 gives the following equations:

Translating

(B1) $$w_i = \frac{p_i}{\mu}(b_i + \alpha_i\eta) + \frac{\alpha_i^c p_i^{1-c}}{\sum \alpha_k^c p_k^{1-c}}\left[1 - \frac{\sum p_k(b_k + \alpha_k\eta)}{\mu}\right].$$

Scaling

$$\text{(B2)} \quad w_i = \frac{p_i}{\mu}(1 + \alpha_i\eta)b_i + \frac{\alpha_i^c[(1 + \alpha_i\eta)p_i]^{1-c}}{\sum \alpha_k^c[(1 + \alpha_k\eta)p_k]^{1-c}}\left[1 - \frac{\sum p_k(1 + \alpha_k\eta)b_k}{\mu}\right].$$

Gorman

$$\text{(B3)} \quad w_i = \frac{p_i}{\mu}\{[1 + (1 - v)\alpha_i\eta]b_i + v\alpha_i\eta\} + \frac{\alpha_i^c\{[1 + (1 - v)\alpha_i\eta]p_i\}^{1-c}}{\sum \alpha_k^c\{[1 + (1 - v)\alpha_k\eta]p_k\}^{1-c}}$$

$$\times \left\{1 - \frac{\sum p_k\{[1 + (1 - v)\alpha_k\eta]b_k + v\alpha_k\eta\}}{\mu}\right\}.$$

Reverse Gorman

$$\text{(B4)} \quad w_i = \frac{p_i}{\mu}[1 + (1 - v)\alpha_i\eta](b_i + v\alpha_i\eta) + \frac{\alpha_i^c\{[1 + (1 - v)\alpha_i\eta]p_i\}^{1-c}}{\sum \alpha_k^c\{[1 + (1 - v)\alpha_k\eta]p_k\}^{1-c}}$$

$$\times \left\{1 - \frac{\sum p_k[1 + (1 - v)\alpha_k\eta](b_k + v\alpha_k\eta)}{\mu}\right\}.$$

Modified Prais–Houthakker

$$\text{(B5)} \quad w_i = \frac{p_i}{\mu}b_i(1 + \alpha_i\eta) + \frac{\alpha_i^c(1 + \alpha_i\eta)p_i^{1-c}}{\sum \alpha_k^c(1 + \alpha_k\eta)p_k^{1-c}}\left[1 - \frac{\sum p_k b_k(1 + \alpha_k\eta)}{\mu}\right].$$

Note: All summations are over k from 1 to 3.

7

Per Capita Time-Series Data

In this chapter we use aggregate time-series data to study several issues in demand system estimation.[1] In Section 1 we investigate alternative demand system specifications and alternative estimation procedures. In Section 2 we examine the effects of pooling data from different countries for demand system estimation.

Section 1 begins by examining alternative specifications of functional form, of dynamic structure, and of stochastic structure. The functional forms we consider are those estimated using household budget data in Chapter 6, Section 2—the LES, QES, BTL, and GTL. The dynamic structures we consider are "dynamic translating" and "dynamic scaling." Both are general procedures in that they do not require the demand system to have a particular functional form, but can be used in conjunction with any complete demand system. The stochastic specifications we consider are the usual "independent" specification in which disturbances in different periods are unrelated, and two specifications permitting first order serial correlation: a "diagonal" specification allowing only one serial correlation parameter for the entire demand system, and a more general "free" specification involving additional serial correlation parameters.

Section 1 concludes with an investigation of estimation procedures that differ in their treatment of the first observation. We estimate each demand system in two ways: using a generalized first difference procedure that excludes the first observation, and using a maximum likelihood procedure that includes it. Our estimates in Section 1 are based on per capita time-series data for the U.S. for the period 1947 through 1983.

In Section 2 we explore the possibility of pooling data from different countries in the estimation of demand systems. Since pooling is most plausible for countries at similar stages of development, we use per capita time series data from three advanced industrial countries: Belgium, the U.K., and the U.S. A nonparametric revealed preference test indicates that the data from these three countries could not have arisen from maximizing a single static nonstochastic utility function. However, pooling may still be possible if there are some short-run or long-run differences in utility

---

[1] This chapter is drawn in part from Pollak and Wales [1987, 1992].

functions among countries. We estimate the QES under alternative specifications that permit such differences in an attempt to confirm the validity of pooling. We use purchasing power parities to transform the data before estimation.

1. ALTERNATIVE DEMAND SPECIFICATIONS AND ESTIMATION METHODS

The four demand systems we estimate are forms of the QES or the GTL. Since these were presented in Chapter 6, Section 2 we do not reproduce them here. The QES in share form is given by Eq. (5) in Chapter 6 and the GTL by Eq. (4). The LES and BTL are special cases of these.

1.1. Dynamic Structure

We estimate the usual static model, in which all demand system parameters remain fixed, and two dynamic specifications that permit some demand system parameters to vary with past consumption. As discussed in Chapter 4, the two dynamic specifications, dynamic translating and dynamic scaling, are general procedures for allowing systematic parameter variation in empirical demand analysis. That is, either specification can be used in conjunction with any original demand system, not just with a restricted class of functional forms. We describe both specifications as modifications of an original class of demand systems, $\{x_i = \bar{h}^i(P, \mu),$ $i = 1, \ldots, n\}$. We assume these original demand systems are "theoretically plausible" (i.e., they can be derived from "well-behaved" preferences), and we denote the corresponding direct utility function by $\bar{U}(X)$ and the indirect utility function by $\bar{\Psi}(P, \mu)$.

Our dynamic translating procedure varies according to whether the original demand system contains constant terms; when it does, we assume that they depend on past consumption. In principle they may depend on any variables representing past consumption (e.g., a weighted average of all past consumption or the highest level of consumption attained in the past), but we assume that they depend only on the previous period's consumption. Denoting the constant terms by d's, we write $d_i = D^i(x_{it-1})$. Log linear dynamic translating, the specification we estimate, is given by

$$(1) \qquad D^i(x_{it-1}) = d_i^* x_{it-1}^{\gamma_i}$$

and adds n parameters to the modified system.

When the original demand system does not contain constant terms, our dynamic translating procedure replaces the original system by

$$(2) \qquad h^i(P, \mu) = d_i + \bar{h}^i(P, \mu - \sum p_k d_k).$$

This introduces n constant terms into the demand system, and we assume they depend on past consumption, as in the preceding case.[2] If the original demand system is theoretically plausible, then the modified system is also, at least for d's close to zero. The modified system satisfies the first order conditions corresponding to the indirect utility function $\Psi(P, \mu) = \bar{\psi}(P, \mu - \sum p_k d_k)$.

Dynamic scaling replaces the original demand system by

$$(3) \qquad\qquad h^i(P, \mu) = m_i \bar{h}^i(p_1 m_1, \ldots, p_n m_n, \mu),$$

where the m's are scaling parameters that depend on the previous period's consumption, $m_i = M^i(x_{it-1})$.[3] If the original demand system is theoretically plausible, then the modified system is also, at least for m's close to one. The modified system satisfies the first order conditions corresponding to the indirect utility function $\Psi(P, \mu) = \bar{\Psi}(p_1 m_1, \ldots, p_n m_n, \mu)$ and the direct utility function $U(X) = \bar{U}(x_1/m_1, \ldots, x_n/m_n)$. Loosely speaking, we might interpret x_i/m_i as a measure of x_i in "efficiency units" rather than physical units.

Log linear dynamic scaling, the specification we estimate, is given by

$$(4) \qquad\qquad M^i(x_{it-1}) = x_{it-1}^{\gamma i}.$$

This specification guarantees that the implied value of m_i will be positive, as is required on theoretical grounds. It adds n parameters to the original demand system.

Specifications allowing demand system parameters to depend on past consumption are formally analogous to those allowing them to depend on demographic variables. Thus, dynamic translating is analogous to demographic translating and dynamic scaling to demographic scaling. Although we do not consider more general dynamic specifications here, all the demographic specifications discussed in Chapter 6, Section 3 have obvious dynamic counterparts.

1.2. Stochastic Structure

We estimate three stochastic specifications, one in which disturbances in different periods are independent and two others that allow first order serial correlation: the "diagonal" autocorrelation specification involves only one serial correlation parameter, while the "free" specification involves $(n-1)^2$ independent serial correlation parameters.

All of our stochastic specifications postulate an additive disturbance term on the share equations. We use the same notation as in Chapter 5

---

[2] A more general formulation of translating would replace (1) by $D^i(x_{it-1}) = c_i + d_i^* x_{it-1}^{\gamma i}$. Given the computational difficulties in converging our simpler models, it is unlikely that this more general specification would converge.

[3] A more general formulation would allow m_i to depend on consumption in all previous periods.

and write

$$(5) \qquad w_t = \omega(z_t, \beta) + \tilde{u}_t,$$

where $\tilde{u}_t = (u_{t1}, \ldots, u_{tn})'$, and $E(\tilde{u}_t) = 0$ for all t.

The free stochastic specification assumes that $\tilde{u}_t$ follows a general first order autoregressive process

$$(6) \qquad \tilde{u}_t = \tilde{R}\tilde{u}_{t-1} + \tilde{e}_t,$$

where $\tilde{R}$ is an n × n matrix of autoregressive parameters and $\tilde{e}_t$ is an n × 1 vector of disturbances with $E(\tilde{e}_t) = 0$ and $E(\tilde{e}_t\tilde{e}_t') = \Omega$ for all t.[4] The diagonal specification is a special case in which $\tilde{R}$ is diagonal; the independent specification corresponds to $\tilde{R} = 0$. Although the diagonal specification might appear to allow n independent serial correlation parameters, it allows only one (see Chapter 5). As usual we can estimate the demand systems after dropping one equation. With three goods, we can identify four independent transformations of the parameters of the $\tilde{R}$ matrix. We denote the 2 × 2 transformed $\tilde{R}$ matrix by R.[5]

1.3. Data and Estimation

Our estimates are based on annual U.S. per capita data for the years 1948–1983. We use three broad commodity groups constructed from the national product accounts: "food," "clothing," and a "miscellaneous" category that excludes shelter, consumer durables, and nondurables that seem closely related to them.[6] Excluding shelter and durables is justified if our three included commodity groups are separable from the excluded ones. The alternative of including the flow of services of shelter and consumer durables would require time series data representing these service flows and the corresponding implicit rents, and would require that consumption of these services be in equilibrium given their implicit rents.[7] Our three included commodity groups account for about half of "personal consumption expenditure." Precise definitions of our commodity groups are given in Appendix A.

---

[4] The only paper we know that estimates demand systems using a free R* matrix is Anderson and Blundell [1982]. They estimate a general dynamic short-run model, using a generalized first difference procedure.

[5] Recall from Chapter 5 that R is obtained from $\tilde{R}$ by first subtracting the last column of $\tilde{R}$ from each of the other columns of $\tilde{R}$, and then deleting the last row and column.

[6] The treatment of shelter is problematic. Although national product accounts purport to measure the flow of housing services, we have excluded shelter because we are skeptical about these data. Howe, Pollak, and Wales [1979] report estimates based on four commodity groups, including shelter.

[7] As an example of this approach see Diewert [1974a]. It is difficult to think of assumptions that justify ignoring durability and treating reported purchases of durables as current consumption flows.

There is evidence, especially strong in the single-equation context, that parameter estimates and hypothesis tests can be sensitive to omission of the first observation. The literature suggests that this is especially likely to be true in small samples when the independent variables exhibit time trends.[8,9] To investigate this phenomenon in the demand system context, we use two alternative estimation procedures discussed in Chapter 5: the "generalized first difference" procedure, which drops the first observation, and the "maximum likelihood" procedure, which does not. Because the generalized first difference procedure is asymptotically equivalent to maximum likelihood, however, the two procedures yield different results only in "small" samples.

1.4. Results

We report results for 30 different models: (i) three dynamic structures (static, dynamic translating, dynamic scaling), (ii) four functional forms (LES, QES, BTL, GTL), and (iii) three stochastic specifications ($R = 0$, R

Table 1 Number of Independent Parameters

	LES	QES	BTL	GTL
Static				
$R = 0$	5	8	8	11
R diagonal	6	9	9	12
R free	9	12	12	15
Dynamic translating				
$R = 0$	8	11	+	14
R diagonal	9	12	+	15
R free	12	15	+	18
Dynamic scaling				
$R = 0$	8	11	11	14
R diagonal	9	12	12	15
R free	12	15	15	18

Notes:
1. We do not include the disturbance covariance parameters in the count.
2. The + denotes the fact that translating can be applied only to systems containing translating parameters; when translating parameters are introduced into the BTL it becomes the GTL.

---

[8] In our sample the normalized prices for food and clothing exhibit a strong time trend, falling almost monotonically over the period. For the miscellaneous category, normalized prices rise initially until about the middle of the period and then fall.

[9] See, for example, Park and Mitchell [1980] andMaeshiro [1979] for single-equation estimation, and Beach and MacKinnon [1979] for systems estimation.

diagonal, R free).[10] Table 1 shows the number of independent parameters (other than disturbance covariance parameters) in each model. Table 2a presents log likelihood values (aside from an additive constant) for 30 models estimated with the generalized first difference procedure and shows the number of sample price–expenditure situations at which each estimated demand system corresponds to well-behaved preferences.[11] Table 2b presents the corresponding values for the maximum likelihood procedure. The likelihood values reported in Tables 2a and 2b are not comparable because the estimates are based on different procedures and different samples.[12]

The results of Tables 2a and 2b may be analyzed in a number of ways. We start with the most general models and then ask whether dynamic structure or stochastic specification restrictions corresponding to a simpler functional form can be imposed.[13] We begin with the LES–QES estimates; the most general models are the QES with R free, and either dynamic translating or scaling. We find that in none of the four such cases can the restrictions corresponding to the LES be imposed, and thus we conclude that the QES is a significant improvement over the LES. We find that in only one of the four cases—dynamic translating with the ML estimates—can the restrictions corresponding to a single serial correlation coefficient be imposed. This is also the only case in which the static model is accepted, and it is accepted only when R is free; when R is constrained to be diagonal, the static model is rejected. Thus, for three of our four most general models the restrictions corresponding to the LES, to a static model, or to a diagonal R are all rejected. In the fourth case—dynamic translating under ML—we reject the LES and either the static model or R diagonal, although not both.

With regard to dynamic specification, we find little to choose between translating and scaling. In our most general models translating yields a higher likelihood value than scaling with the first differencing estimates, but a lower value with the ML estimates. The first differencing estimates, however, do much better at satisfying regularity conditions than do the ML estimates. The former satisfy regularity conditions at all data points for

---

[10] There are 30 rather than 36 distinct models because (a) for the LES, translating and scaling yield identical models and (b) translating cannot be applied to the BTL without adding translating parameters and when translating parameters are introduced into the BTL it becomes the GTL.

[11] Because our estimates are based on aggregate rather than household data, there is no theoretical presumption that these demand functions correspond to well-behaved preferences.

[12] As shown in Chapter 5, the likelihood function for the maximum likelihood procedure contains two terms that do not appear with the generalized first difference procedure; therefore it is not clear a priori which likelihood function will be larger.

[13] We use the standard chi-square test to determine whether a restriction can be imposed.

Table 2a   Log Likelihood Values: Generalized First Difference Procedure

	LES	QES	BTL	GTL
Static				
R = 0	266.72 (16)	276.11 (0)	282.96 (16)	286.96 (12)
R diagonal	299.38 (35)	302.99 (35)	307.75 (0)	309.48 (0)
R free	306.44 (35)	309.53 (35)	309.53 (35)	316.22 (0)
Dynamic translating				
R = 0	301.42 (20)	306.36 (0)	+	309.645 (14)
R diagonal	306.19 (35)	309.89 (35)	+	312.356 (11)
R free	310.27 (35)	315.19 (35)	+	*
Dynamic scaling				
R = 0	301.42 (20)	305.07 (33)	306.29 (23)	309.751 (10)
R diagonal	306.19 (35)	310.32 (35)	311.11 (22)	312.22 (16)
R free	310.27 (35)	314.82 (35)	314.34 (35)	316.75 (35)

Notes:
1. See Table 1.
2. The numbers in parentheses indicate at how many of the 35 sample price–expenditure situations preferences are quasiconcave.
3. The asterisk indicates failure to converge.

Table 2b   Log Likelihood Values: Maximum Likelihood Procedure

	LES	QES	BTL	GTL
Static				
R = 0	272.24 (15)	281.60 (0)	289.33 (15)	292.82 (11)
R diagonal	291.94 (35)	299.69 (12)	300.99 (15)	302.24 (10)
R free	309.83 (35)	313.29 (35)	312.18 (35)	314.98 (35)
Dynamic translating				
R = 0	310.56 (20)	312.00 (26)	+	315.91 (17)
R diagonal	311.39 (33)	314.03 (0)	+	316.74 (17)
R free	312.49 (35)	317.04 (0)	+	*
Dynamic scaling				
R = 0	310.56 (20)	312.79 (32)	313.26 (23)	315.53 (11)
R diagonal	311.39 (33)	313.74 (35)	315.88 (23)	316.22 (15)
R free	312.49 (35)	318.48 (19)	319.17 (20)	320.31 (20)

Note:
1. See Table 2a.

both translating and scaling, while the latter satisfy them at no data points under translating and at 19 of 35 under scaling.

We conclude then that the QES with either dynamic translating or scaling and a free R matrix is our preferred model. The estimation procedure does not appear to affect this general conclusion.

The results for the BTL–GTL estimates are not as clear-cut as those for the LES–QES estimates. The first problem encountered is that neither estimation procedure converged for the GTL with R free under dynamic translating.[14] Thus we do not report results for this model under either estimation procedure. Our most general model then is the GTL with dynamics incorporated through scaling and with R free. As we will see shortly some of our conclusions regarding the translog form are sensitive to the estimation procedure, and/or to the order of testing. There is, however, one result that is robust wth respect to such considerations: in no case can the restrictions corresponding to the simple null hypothesis of a BTL form be rejected.[15] Thus in the following discussion we consider only the BTL results with dynamic scaling. The acceptability of further restrictions on our most general model—the BTL with R free under dynamic scaling—depends on the estimation procedure and, in one case, on the order of testing. For the ML estimates, the null hypothesis of no serial correlation cannot be rejected, after which the static model is rejected: thus, our preferred model incorporates dynamic scaling but no serial correlation of the residuals. For the generalized first difference method we can follow two different test routes. First we can accept R diagonal and then accept the static model, in which case our preferred model is static with a single correlation coefficient for the residuals. Alternatively we can accept the static model with R free but then reject R diagonal. Thus, using the generalized first difference estimation procedure leads us to conclude that the static model, with R diagonal or free, adequately reflects behavior.

The fact that the two estimation procedures yield different conclusions suggests that it is difficult with this functional form to disentangle the dynamic aspects appearing in the stochastic and nonstochastic parts of the model. In one case we have lagged consumption and no serial correlation of the disturbances, and in the other we have no lagged consumption and serial correlation of the disturbances (R either diagonal or free).

Having considered the LES–QES and BTL–GTL forms separately we now make a few comparisons between them. Because the LES is nested in the GTL, a standard likelihood ratio test of the LES restrictions is appropriate. These restrictions involve setting six parameters to 0, with a corresponding 5% chi-square critical value of 12.6. From Tables 2a and 2b it can be seen that in only 5 of the 18 comparisons is the GTL a significant improvement over the LES, and all of these involve static models. This, together with the fact that the GTL is never a significant improvement

---

[14] It should be recalled that the GTL contains three more parameters than the QES—in this case 18 as opposed to 15 in the most general QES models, and with the ML method three additional covariance parameters must be estimated.

[15] This finding differs from that in Chapter 6 where we found the GTL to be superior to the BTL. The difference probably lies in the greater degree of expenditure variation in the cross-section data used in Chapter 6.

over the BTL, suggests that, at least with this time series sample, the GTL is not a very useful form.

The likelihood ratio test cannot be used to compare the QES and the BTL because they are not nested. They do, however, involve the same number of independent parameters, so we can compare them informally on the basis of their likelihood values. This comparison reveals that in 10 of the 12 cases the BTL has a higher likelihood than the QES.

We consider now the effect of different estimation procedures and different functional forms on estimated marginal budget shares and own-price elasticities. In Table 3 we compare the QES with the BTL, each estimated by the generalized first difference and maximum likelihood methods. The models are estimated with dynamic scaling and R free. A comparison of estimation methods reveals substantial differences in both the pattern over time and in the magnitude of some of the estimates. For example, the marginal budget share for food generally declines over time with the maximum likelihood estimates, but generally increases over time with the first difference method. A priori one would expect this share to decline over time as real income rises. The own-price elasticity estimates for food are about the same with the two methods but those for clothing

Table 3 Marginal Budget Shares and Own-Price Elasticities

	Marginal budget shares			Own-price elasticities		
Year	Food	Clothing	Miscellaneous	Food	Clothing	Miscellaneous
	QES (maximum likelihood)					
1950	.52	.33	.15	−.62	−.40	.07
1960	.48	.27	.25	−.61	−.40	−.25
1970	.47	.25	.28	−.64	−.43	−.34
1980	.48	.23	.30	−.67	−.50	−.41
	QES (first difference)					
1950	.39	.24	.37	−.65	−1.02	−.66
1960	.39	.24	.37	−.65	−1.02	−.65
1970	.41	.20	.39	−.70	−.96	−.65
1980	.47	.13	.41	−.71	−.88	−.63
	BTL (maximum likelihood)					
1950	.52	.28	.20	−.67	−.56	−.04
1960	.50	.26	.24	−.68	−.53	−.23
1970	.47	.25	.27	−.67	−.52	−.35
1980	.47	.23	.30	−.67	−.48	−.43
	BTL (first difference)					
1950	.30	.16	.53	−.74	−.74	−.80
1960	.27	.16	.57	−.71	−.74	−.83
1970	.27	.18	.55	−.63	−.69	−.79
1980	.40	.14	.46	−.57	−.40	−.64

and miscellaneous are considerably higher with the first difference procedure.

A comparison of functional forms using the maximum likelihood procedure reveals estimates that are roughly the same. The pattern over time of the marginal budget shares is the same with the two forms and indeed the magnitudes are comparable. The own-price elasticities are remarkably similar, although they are generally slightly higher for the BTL.[16] A comparison of functional forms using the first difference method reveals substantial differences in both the magnitude and pattern over time of the marginal budget shares and in the magnitude of the own-price elasticity for clothing.

The treatment of the first observation may affect the ranking of functional forms and predicted behavior because some of the roots of the R matrix may be near unity for the generalized first difference method. If this is the case, then the term in the likelihood function that restricts the roots to lie in the unit circle under the maximum likelihood procedure may lead to convergence at quite a different parameter vector. To investigate this possibility, we have calculated the roots of the estimated R matrices for the four cases reported in Table 3. For the QES the largest roots are .67 and .98 for the maximum likelihood and generalized first difference methods, respectively, while for the BTL they are .68 and .99. These substantial differences in the roots of the R matrix with the two estimation methods are consistent with our finding that the two methods, differing only in their treatment of the first observation, can yield different predictions about behavior.

Although aggregate demand behavior need not satisfy regularity conditions even if the behavior of each household does, demand systems that violate regularity conditions may imply implausible responses to price or expenditure changes. Tabulating the number of sample price–expenditure situations at which regularity conditions are satisfied, we find that of a total of 420 possible situations the LES, QES, BTL, and GTL satisfy the conditions at 349, 306, 227, and 175 points, respectively.[17] On these grounds, then, the linear and quadratic expenditure models fare better than the translog models.

Finally we have investigated the dynamic stability of the nonstochastic models using simulations based on the estimated parameters. With prices and expenditure fixed at actual levels for the final year of the sample, all models converge to a steady state.

Although the generalized first difference and maximum likelihood procedures yield different estimates and different rankings of functional

---

[16] The QES maximum likelihood estimates predict positive own-price elasticities for the first four years of the sample.

[17] We exclude dynamic translating models from this comparison because they are not defined for the BTL.

forms in some cases, it is not clear which procedure is better. The maximum likelihood procedure requires strong assumptions about the stationarity of the autoregressive process and the normality of the disturbances. In particular, it assumes that the autoregressive process is stationary, whereas the generalized first difference procedure does not.[18] Hence, if the process generating the disturbances changed shortly before the sample period, then the maximum likelihood procedure involves a specification error, while the generalized first difference procedure does not. Since our sample period (1948–1983) begins shortly after World War II, it is plausible that the process generating the disturbances changed shortly before the beginning of our sample period.[19] The maximum likelihood procedure also assumes that the disturbances have a multivariate normal distribution. The generalized first difference procedure can be interpreted as an iterative Zellner procedure that iterates on the elements of the covariance matrix and R as well as on the structural parameters. Thus interpreted, the generalized first difference procedure does not require normality.

Since our sample size—35 price–expenditure situations—is small, even if we accept stationarity and normality we need not prefer maximum likelihood to generalized first difference estimates.[20] The desirable efficiency properties of maximum likelihood hold only asymptotically. Monte Carlo evidence would be useful, but we know of only one study dealing with equation systems. Maeshiro [1980] investigates the properties of various estimators for a nonsingular two-equation linear system assuming that the disturbances have a diagonal serial correlation matrix and that the contemporaneous covariance matrix is known. From his Monte Carlo analysis he concludes that with small samples and trended explanatory variables it is extremely important to use an estimator, such as maximum likelihood, that takes full account of the first observation. Further, of the estimators he studies, the one corresponding most closely to our generalized first difference procedure frequently performs worst.

Similarly, Monte Carlo evidence from single-equation linear models strongly suggests that when the independent variables are trending, retaining the first observation reduces substantially the root mean square

---

[18] By stationarity we mean that the process has been operating for a long time and that the characteristic roots of R [Chapter 5, Eq. (7)] lie within the unit circle. Without stationarity Eq. (8) of Chapter 5 does not hold and the first observation cannot be used separately in the estimation.

[19] Theil [1971, p. 253] and Poirier [1978] make similar arguments in the single-equation case.

[20] In the single-equation autoregressive case, Park and Mitchell [1979] report a Monte Carlo study using data generated from a stationary normal distribution; they find the iterative Prais–Winston procedure slightly superior to maximum likelihood.

error of parameter estimates.[21] To the extent that these results apply to nonlinear equation systems, maximum likelihood may be superior.

In summary we have investigated alternative functional forms, dynamic structures, and stochastic structures for demand systems. In addition we have investigated alternative estimation procedures differing in their treatment of the first observation. We find the QES to be superior to the LES and the GTL not superior to the BTL. Our preferred QES model incorporates either dynamic translating or scaling and a free R matrix; this conclusion is independent of the method of estimation. For the BTL our preferred model depends on the estimation method. For the maximum likelihood procedure it involves dynamic scaling but no serial correlation of the residuals. For the generalized first difference method it involves a static model with R either diagonal or free. The fact that the two estimation procedures yield different conclusions suggests that it is difficult with this functional form to disentangle the dynamic aspects appearing in the stochastic and deterministic parts of the model.

2. POOLING INTERNATIONAL CONSUMPTION DATA

In this section we report tests of pooling per capita time-series consumption data. Since pooling is most plausible for countries at similar stages of development, we use data from three advanced industrial societies: Belgium, the U.K., and the U.S. A nonparametric revealed preference test shows that the data from these three countries could not have arisen from maximizing a single, static, nonstochastic utility function. Thus we are led to incorporate dynamic elements and/or permanent differences among countries into the utility function. We estimate the QES using various specifications that permit both short-run and long-run differences among countries in an attempt to confirm the validity of pooling.

2.1. Demand System Differences

When countries' demand systems are identical, efficient estimation requires pooling. When countries' demand systems are unrelated, pooling is impossible: knowing one country's demand system parameters provides no information about any other's.[22] Between these extremes of identical

---

[21] See, for example, Beach and MacKinnon [1978], Harvey and McAvinchey [1978], Maeshiro [1979], and Park and Mitchell [1980]. In the single-equation case our generalized first difference procedure corresponds to the iterative Cochrane–Orcutt method. Although we have not presented results for the system analogue of the iterative Prais–Winston method, there is evidence (Park and Mitchell [1979]) that in the single-equation case it is very similar to maximum likelihood in small samples.

[22] We ignore the possibility of efficiency gains from taking account of nonzero disturbance covariances among countries.

and unrelated demand systems is a continuum of cases in which there are gains from pooling. We consider two polar cases permitting pooling and a comprehensive case that includes them both. Our two polar cases are the "permanent difference" specification in which some demand system parameters differ across countries and the "transitory difference" specification in which the underlying parameters are identical in all countries, but each country's short-run consumption pattern depends on its own past consumption.

Although our specifications of demand system differences can be applied to any demand system, we introduce them using the QES. The QES in share form is given by Eq. (5) in Chapter 6 and is not repeated here.

The permanent difference specification postulates that some subset of demand system parameters differs across countries, while the remaining parameters do not. For example, with the QES we might postulate that the b's differ across countries while the a's and c's are the same everywhere or, alternatively, that the a's and b's differ across countries while the c's are the same.[23] Since theoretical considerations do not dictate which parameters differ, the choice must be based on econometric convenience and empirical plausibility. Provided that there is some specified subset of demand system parameters that are identical across countries, efficient estimation requires pooling.[24]

The transitory difference specification postulates that some subset of demand system parameters depends on past consumption while the remaining parameters are constant and identical in all countries. For example, in the QES we might postulate that the b's depend linearly on consumption in the previous period:

$$(7) \qquad\qquad b_{it} = b_i^* + \beta_i x_{it-1}.$$

The transitory difference specification assumes that the underlying parameters, the a's, b*'s, c's, and β's, are the same in all countries. This yields a model in which short-run demand responses to changes in prices and expenditure differ across countries because past consumption patterns differ, although the long-run or steady-state demand behavior is the same in all countries. The transitory difference specification implies a dynamic model in which the consumption pattern corresponding to a particular

---

[23] The parameters that differ need not constitute a "natural" subset: we might postulate that a_1, a_2, and b_1 differ, but that the c's and the remaining a's and b's are the same everywhere.

[24] The usefulness of a particular specification depends on the demand system functional form and the subset of parameters assumed to vary from one country to another. Suppose we have only one budget study from each of several countries and wish to estimate the LES. We can do so if we assume that all countries have identical demand systems, or, using the permanent difference specification, if we assume that the a's vary while the b's are the same in all countries. If we assume that the b's vary while the a's are the same, we cannot identify the b's. Thus, in conjunction with specific assumptions about which parameters vary from one country to another, the permanent difference specification may prove useful in pooling household budget data from different countries.

price–expenditure situation depends on past consumption. Since empirical studies using per capita time-series data from a single country have generally found dynamic specifications significantly superior to static ones, the transitory difference specification is attractive for pooling per capita time-series data.[25] Alternative transitory difference specifications correspond to different assumptions about which demand system parameters depend on past consumption and about the form of this dependence. Using the previous period's consumption to represent past consumption is convenient, but weighted averages of past consumption or the highest previous level are plausible and tractable alternatives.

The permanent difference specification is static—short-run and long-run demand functions coincide. It does, however, allow persistent differences across countries in demand functions, perhaps reflecting underlying differences in climate, in stable or slowly changing demographic characteristics, or in household production technology or tastes. The transitory difference specification is dynamic: countries with different consumption histories exhibit different consumption patterns when they face identical price–expenditure situations; in the long run, however, the consumption patterns of countries facing identical price–expenditure situations will converge. Although the transitory difference specification is compatible with habit formation or interdependent preferences, lagged consumption may be statistically significant in empirical demand analysis because of omitted variables unrelated to taste change; hence, although we find endogenous tastes a plausible hypothesis, we avoid equating the transitory difference specification with endogenous tastes.

The comprehensive specification combines the permanent and transitory difference specifications, allowing both persistent long-run differences in demand and a dynamic structure. For the QES one such comprehensive specification is obtained by modifying the transitory difference specification by assuming the b*'s differ across countries, while still requiring the a's, c's, and β's to be the same in all countries. Alternative comprehensive specifications are obtained by assuming, for example, that only the a's differ across countries, or that only the c's differ, or that both the a's and c's differ.

2.2. Data and Results

2.2.1. Preliminaries

In this section we report tests of pooling per capita consumption data from Belgium, the U.K., and the U.S. for the years 1961–1978. We consider

---

[25] Our transitory difference specification is similar to the Anderson and Blundell [1982] approach (i.e., embedding steady-state solutions in a more general dynamic framework). The transitory difference specification is unlikely to be useful for analyzing household budget data because such data sets seldom report past consumption.

three commodity groups: food, clothing, and a nondurable miscellaneous group that excludes housing and educational expenditures.

Data availability determined our choice of countries. First, we considered only the 16 countries for which Kravis, Heston, and Summers [1978] report purchasing power parity (PPP) data.[26] Second, since pooling is most plausible for countries at similar stages of development, we considered only advanced industrial societies. Third, since pooling requires comparable commodity groups, we considered only countries for which the OECD National Accounts 1961–1978 [1980] reports data from which we could construct comparable food, clothing, and miscellaneous categories. Finally, since we wished to test the appropriateness of pooling, we required a sufficiently long time series for each country to estimate its demand system separately; although consumption data for several countries were available for the period 1970–1978, we considered only countries for which data were available for the full 18-year period.[27]

The demand system we estimate, the QES, is given (in share form) by Eq. (5) in Chapter 6. We obtain a stochastic specification by adding a disturbance term to each share equation. Denoting the $n \times 1$ vector of disturbances $(u_{t1}, \ldots, u_{tn})'$ by $\tilde{u}_t$, we assume that the $\tilde{u}_t$ are independently distributed across countries but that within each country they are correlated over time according to the simple first order scheme

$$(8) \qquad\qquad \tilde{u}_t = \tilde{R}\tilde{u}_{t-1} + \tilde{e}_t,$$

where the $\tilde{e}_t$ are independently normally distributed with covariance matrix $\tilde{\Omega}$ and $\tilde{R}$ is a diagonal matrix with ρ on the main diagonal.[28] We obtain maximum likelihood estimates of the demand system parameters (with b related to past consumption as in (7)) of Ω and of ρ by maximizing the concentrated likelihood function associated with the model using all observations, including the first.[29]

2.2.2. Pooling Three Countries

Using the nonparametric revealed preference test procedure of Afriat [1967], Diewert [1973], and Varian [1982], we find that the data for each country separately could be generated by maximizing a static, nonstochastic utility function, but that the pooled data for all three countries could not be generated by maximizing a single, static, nonstochastic utility function.[30] Thus pooling data from different countries, if it is possible at

---

[26] As an alternative to using published PPP data, in Appendix C we describe a procedure that could be used in some cases to estimate PPPs along with demand system parameters.

[27] A complete description of data sources appears in Appendix B.

[28] We also allowed ρ to differ across countries, but since this had very little effect on the likelihood values, we have not reported those results here.

[29] As in Chapter 5, Ω corresponds to the first $n - 1$ rows and columns of $\tilde{\Omega}$.

[30] We are grateful to Hal Varian for making his programs available to us.

all, requires a specification that permits demand system parameters to differ across countries or to change over time.

An alternative approach to determining whether data from the three countries can be pooled is to estimate a utility function of a particular functional form. We estimate the QES for each country separately and for all three countries combined, allowing neither transitory nor permanent differences among countries, and use a standard likelihood ratio test to determine if the data from all three countries are generated by the same process. Pooling three countries constrains 8 demand system parameters (3 b's, 3 c's, and 2 a's) and 4 covariance parameters (3 elements of Ω and the value of ρ) to be the same in all countries; thus the likelihood ratio test of equality for three countries uses 24 degrees of freedom (two pairwise country comparisons involving 12 parameters for each country). Since our calculated chi-square value is 115.9 while the critical 1% value with 24 degrees of freedom is 43.0, we reject pooling of all three countries, a result consistent with that obtained using Varian's nonparametric test.[31,32] Thus we are led to try a specification that allows for transitory and/or permanent differences among countries.

We allow for transitory differences among countries by estimating the QES with dynamic translating. More specifically, as in Eq. (7) we assume that the b's for each good are linearly related to consumption of that good in the previous period, a specification we have called "linear dynamic translating" in Chapter 4.[33] With this dynamic specification the demand system for each country contains 4n independent parameters, including a serial correlation coefficient. We use the standard likelihood ratio test to determine whether the demand system parameters are the same in all three countries. Our calculated chi-square value is 119.0 while the critical value is 50.9. Thus we reject pooling data for all three countries

---

[31] We use the 1% critical values in an attempt to offset the small sample bias of the likelihood ratio test. Using the LES in a Monte Carlo experiment, Wales [1984] tested the (true) null hypothesis of no difference in parameter values for two data sets of size 20 at the 1% level and rejected the null hypothesis about 4% of the time. Thus our best guess is that using the 1% critical values corresponds to a Type I error of about 4%.

[32] Although the likelihood ratio test rejects pooling of all three countries using a static specification without permanent differences, the estimated pooled static QES does satisfy regularity conditions at all price–expenditure situations in the sample. This result does not contradict the nonparametric test result, but it does illustrate an important difference between the nonparametric and estimation approaches. The nonparametric approach examines regularity conditions using observed prices and observed quantities, while the demand system estimation approach uses observed prices and predicted quantities. That is, the estimation approach examines whether the estimated demand system is consistent with regularity conditions at each observed price–expenditure situation. The stringency of this procedure depends on the particular functional form: for example, if in the data all goods have nonnegative shares, then an estimated demand system corresponding to a Cobb–Douglas utility function must satisfy regularity conditions.

[33] We also estimated our model using the "dynamic scaling" procedure discussed in Chapter 4, but the results were so similar to those of dynamic translating that we have not reported them here.

using the QES with dynamic translating. The nonparametric approach is applicable only to static specifications. It cannot be used to determine whether the pooled data for all three countries could have been generated by maximizing a single dynamic utility function with transitory differences introduced through dynamic translating. Thus, continuing with our parametric approach, we now turn to a specification that allows for permanent differences among countries.

Comprehensive specifications allow both a dynamic structure within countries and long-run differences among countries. We next consider four comprehensive specifications, each of which allows one subset of parameters (either the a's, b's, c's, or β's) to vary among countries but assumes that the remaining parameters are identical in all three countries.[34] We see no a priori reason to prefer one of these specifications to the others. Letting only the b's, c's, or β's vary involves 24 restrictions and a critical chi-square value of 43.0, while our calculated values are 52.9, 82.2, and 67.8, respectively. Letting the a's vary involves 26 restrictions and a critical value of 45.6, while our calculated value is 67.0. Thus, even using the comprehensive specifications, the likelihood ratio test rejects the pooling of all three countries.

2.2.3. Pooling Two Countries

The results for pooling pairs of countries are more complex. Nonparametric tests show that the data from one pair of countries (U.K.–U.S.) are consistent with maximization of a static, nonstochastic utility function, while data from the other two pairs of countries (Belgium–U.K., Belgium–U.S.) are not. Estimating the QES for each of the three pairs of countries without allowing either permanent or transitory differences, we find that the likelihood ratio test rejects pooling for all three pairs. This result appears in the first row of Table 4, which records the calculated chi-square values used in the likelihood ratio tests. Although pooling is rejected for the U.K.–U.S. pair, the rejection is far less decisive than it is for the other two pairs of countries.

Since the nonparametric test shows that the U.K.–U.S. data are consistent with maximizing some utility function, the rejection of pooling with the QES could be interpreted as evidence of functional form misspecification. The rejection, however, might also be interpreted as a reflection of dynamic misspecification. Suppose, for example, that the data were generated by a dynamic QES; for a particular set of price–expenditure

---

[34] The only other reference in the literature of which we are aware that involves this type of testing is Goldberger and Gamaletsos [1970]. Discussing the possibility of pooling data for 13 OECD countries, these authors conclude "computations not reported here indicate rejection of the hypothesis that all parameters are fixed across countries, but leave open the possibility that subsets of parameters possess that invariance property" (p. 381).

Table 4   Chi-Square Values for Pooling Pairs of Countries

	DOF	B.–U.K.	U.K.–U.S.	B.–U.S.
Static utility function	12	82.1	35.2	65.6
Dynamic utility function				
No parameters differing	15	76.1	77.4	68.0
b's differing	12	35.2	23.9	23.9
c's differing	12	43.5	36.2	45.5
β's differing	12	40.2	52.9	29.1
a's differing	13	35.6	44.8	38.4

Note:
1. DOF denotes degrees of freedom for the likelihood ratio test. The 1% critical values
   for 12, 13, and 15 degrees of freedom are 26.7, 27.7, and 30.6, respectively.

situations, the price–quantity data might pass the nonparametric consistency test while failing the likelihood ratio test for pooling based on the estimated but misspecified static QES. Alternatively, the rejection might be interpreted as a reflection of a failure to specify appropriately permanent differences among countries. Although one's first instinct is that Occam's razor argues for treating constant tastes as a maintained hypothesis, this is not the case. Such an argument depends on the implicit assumption that constant tastes is the simpler hypothesis—regardless of the complexity of the static utility function required to rationalize observed behavior.

As in the case of pooling three countries we next consider a specification that allows transitory but not permanent differences between countries. We estimate the QES with dynamic translating for each country separately and for each pair of countries and use likelihood ratio tests to assess the validity of pooling. For each of these tests, the calculated chi-square value, which appears in the second row of Table 4, is over twice as large as the corresponding critical value. Thus we reject pooling pairs of countries for the QES with dynamic translating and conclude that, with our data and demand system, allowance must be made for permanent differences between countries.

As in the three-country case, our comprehensive specifications allow for both a dynamic structure and long-run differences between countries. The calculated chi-square values for the four comprehensive specifications, which appear in rows 3 through 6 of Table 4, indicate that we reject pooling in 10 of 12 cases. In particular, we do not reject pooling for the Belgium–U.S. and the U.K.–U.S. pairwise comparisons when the b's are allowed to differ. We reject pooling in all other cases. That is, we reject pooling when parameter sets other than the b's are allowed to vary, and we reject pooling of the Belgium–U.K. pair when the b's are allowed to differ.

To predict elasticities and marginal budget shares we use the pooled Belgium–U.S. estimates for Belgium and the pooled U.K.–U.S. estimates

Table 5 Marginal Budget Shares and Own-Price Elasticities: Pooled Data

Marginal budget shares			Own-price elasticities		
Food	Clothing	Miscellaneous	Food	Clothing	Miscellaneous
Belgium (based on pooled Belgium–U.S. data)					
.38	.18	.44	− .42	− .28	− .54
.35	.18	.47	− .47	− .45	− .67
.31	.18	.51	− .50	− .60	− .78
United Kingdom (based on pooled U.K.–U.S. data)					
.15	.24	.61	− .10	− .54	− .78
.13	.21	.66	+ .09	− .46	− .65
.17	.18	.65	− .07	− .45	− .65
United States (based on pooled U.K.–U.S. data)					
.26	.19	.55	− .79	− .81	− 1.05
.23	.19	.57	− .68	− .72	− .95
.27	.17	.57	− .66	− .67	− .88
United States (based on pooled Belgium–U.S. data)					
.25	.18	.57	− .64	− .71	− .92
.21	.18	.61	− .60	− .73	− .93
.18	.17	.65	− .56	− .75	− .93

Notes:
1. Demand system parameters were estimated using pooled data from two countries; marginal budget shares and elasticities were evaluated at the price–expenditure situation of the designated country for the first, middle, and last year of the sample.
2. The demand system estimated using pooled Belgium–U.S. data satisfies regularity conditions at all 17 Belgian price–expenditure situations in our sample. Similarly, the estimates using pooled U.K.–U.S. data satisfy regularity conditions at all 17 U.S. sample points, and the estimates using Belgium–U.S. data satisfy regularity conditions at all 17 U.S. sample points. The demand system estimated using pooled U.K.–U.S. data, on the other hand, violates regularity conditions at all 17 U.K. price–expenditure situations in our sample.

for the U.K., with the b's allowed to vary. These estimates are presumably more efficient than those based on individual country data alone.[35] For the U.S., with two conflicting sets of pooled estimates, there is no conventional solution, and we present estimates based on both. In Table 5 we report estimated marginal budget shares and own-price elasticities at three price–expenditure situations for each country, those corresponding to the first, middle, and last years of our sample. All estimated marginal budget shares and elasticities appear reasonable a priori except for the positive own-price elasticity for food in the U.K. in the middle sample period.[36] For the U.S. the two sets of estimates are very similar at the middle sample point, but in several cases the trends over the sample period differ.

---

[35]When countries are treated separately we estimate 15 parameters using 18 observations (i.e., 18 years of price and quantity data on three goods), while when countries are treated in pairs with the b's varying we estimate 18 parameters using 36 observations (on three goods).

[36]Although not shown in the table, this positive elasticity does not differ significantly from 0 at the 5% level.

Table 5 also reports the number of sample price–expenditure situations at which the estimated demand system parameters correspond to well-behaved preferences. For the U.S. and for Belgium, these regularity conditions are satisfied at all 17 sample points for both sets of estimates; for the U.K. they are not satisfied at any sample points.[37]

Finally we report on our tests of functional form and dynamic structure. We find the QES functional form significantly superior to the LES, the null hypothesis being that $c_1 = c_2 = c_3 = 0$; the calculated chi-square values for the Belgium–U.S. and U.K.–U.S. pairs are 29.3 and 19.9, while the critical 1% value is 6.6, thus indicating significantly nonlinear expenditure responses. For dynamic structure our findings are mixed. To test whether the short-run and long-run demand functions differ, the null hypothesis is $\beta_1 = \beta_2 = \beta_3 = 0$. The calculated chi-square values for the Belgium–U.S. and U.K.–U.S. pairs are 2.6 and 22.5, while the 1% critical value is 6.6. Thus, using the pooled estimates, we reject the hypothesis that short-run and long-run demands are the same for the U.K., but we do not reject it for Belgium; for the U.S. we have conflicting results.

In summary we find that even under circumstances favorable to pooling—using countries at the same general stage of development—pooling is not generally acceptable. Nonparametric revealed preference tests show that the pooled data from Belgium, the U.K., and the U.S. could not have been generated by maximizing a single, static, nonstochastic utility function. Our attempts to pool data using a parametric approach based on the QES, allowing for both permanent differences among countries and a dynamic structure within countries, were generally unsuccessful.

We have considered models in which no parameters vary across countries and those in which only one set varies; for most parameter sets and combinations of countries, we rejected pooling. We find this result disappointing since our specifications were designed to meet the legitimate objection that it is implausible to assume that all countries have identical short-run and long-run demand behavior. Furthermore, the three countries we analyzed represented a case favorable to pooling. There is no reason to believe we would have done better pooling data from Canada, Germany, and the Netherlands, for example, and some reason to believe we would have done worse pooling data from France, India, and Kenya.[38]

---

[37]We calculate regularity conditions at 17 rather than 18 sample points because, in our dynamic models, calculating regularity conditions requires the previous period's consumption pattern.

[38]Our modeling of demand differences between countries may be too simple; it is possible that we would do better allowing more than one parameter set to vary. Thus, we could allow two parameter sets to vary across countries, say the b's and c's, while requiring the β's and a's to be the same. We have estimated specifications of this type for all six pairs of parameter sets and all three pairs of countries. Although detailed results are not reported here, we find that pooling is permissible in 11 of the 18 cases. Pooling data for all three countries is, however, never permissible.

Our results suggest that researchers should be cautious about pooling data from different countries for demand system estimation.

APPENDIX A: U.S. DATA, 1948–1983

Constant (1972) and current dollar expenditures on the various categories of goods for the period 1948–1975 were obtained from Tables 2.7 and 2.6, respectively, of *The Survey of Current Business*, July 1978. In terms of the categories, we defined our three commodity groups as follows (numbers in parentheses correspond to those in Table 2.7):

A. Food
  1. Food (17)
B. Clothing
  1. Clothing and shoes (23)
  2. Shoe cleaning and repair (61)
  3. Cleaning, laundering, dyeing, etc. (62)
C. Miscellaneous
  1. Toilet articles and preparations (31)
  2. Tobacco products (30)
  3. Drug preparations and sundries (34)
  4. Nondurable toys and sports supplies (35)
  5. Domestic service (48)
  6. Barbershops, beauty parlors, and baths (63)
  7. Medical care services (64)
  8. Admissions to specified spectator amusements (69)

Constant (1972) and current dollar expenditures on the various categories of goods for the period 1976–1983 were obtained from Tables 2.4 and 2.5, respectively, of *The Survey of Current Business*, July 1984. In terms of the categories, we defined our three groups as follows (numbers in parentheses correspond to those in Table 2.5):

A. Food
  1. Food (20)
B. Clothing
  1. Clothing and shoes (27)
  2. Cleaning, storage, and repair of clothing and shoes (68)
C. Miscellaneous
  1. Toilet articles and preparations (35)
  2. Tobacco products (34)
  3. Drug preparations and sundries (38)
  4. Nondurable toys and sports supplies (39)
  5. Domestic service (54)
  6. Barbershops, beauty parlors, and baths (69)

7. Medical care services (71)

8. Admissions to specified spectator amusements (83)

Per capita consumption of each good was calculated by dividing annual expenditure in 1972 dollars by population. The population figures are "resident population" in the U.S. and are taken from Table 2 of *The Statistical Abstract of the U.S., 1976*, for the period 1948–1972 and from *Current Population Reports*, Series P 25, No. 983, June 1984 for the period 1973–1983. Price indexes were determined by dividing current dollar expenditure by constant dollar expenditure for each of the three categories.

APPENDIX B: OECD DATA, 1961–1978

Consumer Expenditure Data

Expenditure in current prices and constant prices on the various categories of goods were obtained from Tables 4a and 5b of the *OECD National Accounts 1961–1978*, Volume 2. In terms of the categories, we defined our three broad commodity groups as follows (numbers in parentheses correspond to those in Tables 5a and 5b):

A. Food
   1. Food (1-1)
   2. Nonalcoholic beverages (1-2)
   3. Alcoholic beverages (1-3)
B. Clothing
   1. Clothing and footwear (2)
C. Miscellaneous
   1. Other recreation, etc. (7-2)
   2. Miscellaneous goods and services (8)
   3. Tobacco (1-4)

Per capita expenditure on each commodity group was calculated by dividing annual expenditure by mid-year population. The population data for each country are from the *U.N. Demographic Yearbook, 1979*. Price indexes were obtained by dividing current price expenditure by constant price expenditure; we normalized the price index series to unity in 1970 by dividing each series by its 1970 value.

Purchasing Power Parity Data

The PPP data are from Tables 5.1 and 5.15 of Kravis, Heston, and Summers [1978]. In terms of the categories our three groups were defined as follows:

A. Food
   1. Food

     2. Beverages
  B. Clothing
     1. Clothing
  C. Miscellaneous
     1. Recreation
     2. Other expenditures (personal care and miscellaneous services)
     3. Tobacco

The resulting categories appear very similar to those constructed from the expenditure data. We aggregated these PPP data to our commodity group levels by combining the country-weighted PPPs from the tables using the corresponding country expenditures as weights to give country-weighted PPPs at the commodity group level. The PPPs we used are the geometric means of these country-weighted PPPs for each commodity group. To obtain comparable quantities, we divided the normalized Belgian and U.K. quantity data for food, clothing, and miscellaneous by the corresponding PPPs; comparable price series were obtain by multiplying the normalized price series by the PPPs.

APPENDIX C: ESTIMATING PURCHASING POWER PARITIES

When reported PPPs are not available, it is sometimes possible to estimate the required transformation factors along with the demand system parameters.[39] The PPPs—one for each good in each country, except the base country for which they are unity—enter the demand system by multiplying the prices and dividing the quantities.

    We attempted to estimate PPPs for the U.K.–U.S. and Belgium–U.S. pairs analyzed above, using the QES in which the b's differ between countries. Because the PPPs and the b's enter the QES demand equations in very similar ways, separating their effects is bound to be difficult. We found that the likelihood function was extremely flat with respect to the PPPs and b's, and we were unable to converge the model. Indeed if the LES rather than the QES were being estimated the model would not be identified. This suggests that a specification allowing parameters other than the b's to vary might perform better. Since our data do not accept pooling when other single-parameter sets are allowed to differ, we have estimated specifications in which two sets of parameters are allowed to differ. For example, when we allow the a's and c's to vary, then pooling the U.S. and U.K. data is acceptable; with this model we have estimated PPP values for the U.K. of .46, .93, and .32 for food, clothing, and miscellaneous, respectively. Kravis, Heston, and Summers report values of .35, .33, and

---

[39]This technique for estimating unobserved transformation factors is analogous to demographic scaling.

.35, while the exchange rate in 1970 was .4174. Although the estimates for food and miscellaneous are fairly close to the reported values, the corresponding standard errors are .42, 1.6, and 3.9, so the point estimates are of limited interest. Nevertheless this illustrates a technique that might allow the pooling of other data sets in the absence of PPP information.

References

Afriat, S. N. 1967. "The Construction of Utility Functions from Expenditure Data." *International Economic Review* 8:1 (February): 67–77.

Anderson, Gordon J., and Richard W. Blundell. 1982. "Estimation and Hypothesis Testing in Dynamic Singular Equation Systems." *Econometrica* 50:6 (November): 1559–1572.

Anderson, Gordon J., and Richard W. Blundell. 1983. "Testing Restrictions in a Flexible Dynamic Demand System: An Application to Consumers' Expenditure in Canada." *Review of Economic Studies* 50:162 (July): 397–410.

Arrow, Kenneth J. 1963. *Social Choice and Individual Values*, 2nd ed. Cowles Foundation for Research in Economics. New York: Wiley (1st ed., 1951).

Arrow, Kenneth J. 1971. "A Utilitarian Approach to the Concept of Equality in Public Expenditures." *Quarterly Journal of Economics* 85:3 (August): 409–415.

Arrow, Kenneth J., Hollis B. Chenery, Bagicha S. Minhas, and Robert M. Solow. 1961. "Capital-Labor Substitution and Economic Efficiency." *Review of Economics and Statistics* 43:3 (August): 225–250.

Avery, Robert B. 1977. "Error Components and Seemingly Unrelated Regressions." *Econometrica* 45:1 (January): 199–210.

Baltagi, Badi H. 1980. "On Seemingly Unrelated Regressions with Error Components." *Econometrica* 48:6 (September): 1547–1551.

Barten, Anton P. 1964. "Family Composition, Prices and Expenditure Patterns." In *Econometric Analysis for National Economic Planning: 16th Symposium of the Colston Society*, edited by Peter Hart, Gordon Mills, and John K. Whitaker, pp. 277–292. London: Butterworth.

Barten, Anton P. 1969. "Maximum Likelihood Estimates of a Complete System of Demand Equations." *European Economic Review* 1:1 (Fall): 7–73.

Beach, Charles M., and James G. MacKinnon. 1978. "A Maximum Likelihood Procedure for Regression with Autocorrelated Errors." *Econometrica* 46:1 (January): 51–58.

Beach, Charles M., and James G. MacKinnon. 1979. "A Maximum Likelihood Estimation of Singular Equation Systems with Autoregressive Disturbances." *International Economic Review* 20:2 (June): 459–464.

Becker, Gary S. 1965. "A Theory of the Allocation of Time." *Economic Journal* 75:299 (September): 493–517

Becker, Gary S., and Kevin M. Murphy. 1988. "A Theory of Rational Addiction." *Journal of Political Economy* 96:4 (August): 675–700.

Bergson (Burk), Abram. 1936. "Real Income, Expenditure Proportionality, and

Frisch's 'New Methods of Measuring Marginal Utility'." *Review of Economic Studies* 4 (October): 33–52.

Berndt, Ernst R., Masako N. Darrough, and W. Erwin Diewert. 1977. "Flexible Functional Forms and Expenditure Distributions: An Application to Canadian Consumer Demand Functions." *International Economic Review* 18:3 (October): 651–675.

Berndt, Ernst R., and N. Eugene Savin. 1975. "Estimation and Hypothesis Testing in Singular Equation Systems with Autoregressive Disturbances." *Econometrica* 43:5–6 (September–November): 937–958.

Blackorby, Charles, and David Donaldson. 1989. "Adult-Equivalence Scales and the Economic Implementation of Interpersonal Comparisons of Well-Being." University of British Columbia Discussion Paper 88–27, revised.

Blackorby, Charles, Daniel Primont, and R. Robert Russell. 1978. *Duality, Separability, and Functional Structure: Theory and Economic Applications.* New York: North-Holland.

Blair, Douglas H., and Robert A. Pollak. 1983. "Rational Collective Choice." *Scientific American* 249:16 (August): 88–95.

Blundell, Richard. 1988. "Consumer Behaviour: Theory and Empirical Evidence—A Survey." *Economic Journal* 98:389 (March): 16–65.

Boyce, Richard. 1975. "Estimation of Dynamic Gorman Polar Form Utility Functions." *Annals of Economic and Social Measurement* 4:1 (Winter): 103–116.

Boyer, Marcel. 1983. "Rational Demand and Expectations Patterns under Habit Formation." *Journal of Economic Theory* 31:1 (October): 27–53.

Brown, Murray, and Dale Heien. 1972. "The *S*-Branch Utility Tree: A Generalization of the Linear Expenditure System." *Econometrica* 40:4 (July): 737–747.

Case, Anne C. 1991. "Spatial Patterns in Household Demand." *Econometrica* 59:4 (July): 953–965.

Christensen, Laurits R., Dale W. Jorgenson, and Lawrence J. Lau. 1975. "Transcendental Logarithmic Utility Functions." *American Economic Review* 65:3 (June): 367–383.

Cooter, Robert, and Peter Rappoport. 1984. "Were the Ordinalists Wrong About Welfare Economics?" *Journal of Economic Literature* 22:2 (June): 507–530.

Darrough, Masako N., Robert A. Pollak, and Terence J. Wales. 1983. "Dynamic and Stochastic Structure: An Analysis of Three Time Series of Household Budget Studies." *Review of Economics and Statistics* 65:2 (May): 274–281.

David, Martin H. 1962. *Family Composition and Consumption.* Amsterdam: North-Holland.

Davidson, Russell, and James G. MacKinnon. 1981. "Several Tests for Model Specification in the Presence of Alternative Hypotheses." *Econometrica* 49:3 (May): 781–794.

Deaton, Angus. 1981. "Introduction to part one." In *Essays in the Theory and Measurement of Consumer Behaviour in Honour of Sir Richard Stone,* edited by Angus Deaton, pp. 7–29. Cambridge: Cambridge University Press.

Deaton, Angus, and John Muellbauer. 1980a. "An Almost Ideal Demand System." *American Economic Review* 70:3 (June): 312–326.

Deaton, Angus, and John Muellbauer. 1980b. *Economics and Consumer Behavior.* Cambridge: Cambridge University Press.

Deaton, Angus S., and John Muellbauer. 1986. "On Measuring Child Costs: With Applications to Poor Countries." *Journal of Political Economy* 94:4 (August): 720–744.

Debreu, Gerard. 1960. "Topological Methods in Cardinal Utility Theory." In *Mathematical Methods in the Social Sciences, 1959*, edited by Kenneth J. Arrow, Samuel Karlin, and Patrick Suppes, pp. 16–26. Stanford: Stanford University Press.

Diewert, W. Erwin. 1973. "Afriat and Revealed Preference Theory." *Review of Economic Studies* 40:3 (July): 419–425.

Diewert, W. Erwin. 1974a. "Intertemporal Consumer Theory and the Demand for Durables." *Econometrica* 42:3 (May): 497–516.

Diewert, W. Erwin. 1974b. "Applications of Duality Theory." In *Frontiers of Quantitative Economics, Vol. 2*, edited by Michael D. Intriligator and David A. Kendrick, pp. 106–171. Amsterdam: North-Holland.

Diewert, W. Erwin. 1981. "The Economic Theory of Index Numbers: A Survey." In *Essays in the Theory and Measurement of Consumer Behaviour in Honour of Sir Richard Stone*, edited by Angus Deaton, pp. 163–208. Cambridge: Cambridge University Press.

Diewert, W. Erwin. 1982. "Duality Approaches to Microeconomic Theory." In *Handbook of Mathematical Economics, Vol. 2*, edited by Kenneth J. Arrow and Michael D. Intriligator, pp. 535–599. Amsterdam: North-Holland.

Diewert, W. Erwin. 1983. "The Theory of the Cost-of-Living Index and the Measurement of Welfare Change." In *Price Level Measurement*, edited by W. Erwin Diewert and Claude Montmarquette, pp. 163–233. Ottawa: Statistics Canada.

Diewert, W. Erwin, and Terence J. Wales. 1987. "Flexible Functional Forms and Global Curvature Conditions." *Econometrica* 55:1 (January): 43–68.

Diewert, W. Erwin, and Terence J. Wales. 1988. "A Normalized Quadratic Semi-flexible Functional Form." *Journal of Econometrics* 37:3 (March): 327–342.

Duesenberry, James S. 1949. *Income, Saving, and the Theory of Consumer Behavior.* Cambridge: Harvard University Press.

Easterlin, Richard A. 1973. "Relative Economic Status and the American Fertility Swing." In *Family Economic Behavior: Problems and Prospects*, edited by Eleanor B. Sheldon, pp. 170–223. Philadelphia: J. B. Lippincott.

Easterlin, Richard A., 1976. "The Conflict between Aspirations and Resources." *Population and Development Review* 2:3/4 (September/December): 417–425.

Easterlin, Richard A., Robert A. Pollak, and Michael L. Wachter. 1980. "Towards a More General Model of Fertility Determination: Endogenous Prefer-ences and Natural Fertility." In *Population and Economic Change in Less Developed Countries*, edited by Richard A. Easterlin, pp. 81–135. (Universities-National Bureau of Economic Research Conference Series). Chicago: University of Chicago Press.

El-Safty, Ahmad E. 1976a. "Adaptive Behavior, Demand and Preferences." *Journal of Economic Theory* 13:2 (October): 298–318.

El-Safty, Ahmad E. 1976b. "Adaptive Behavior and the Existence of Weizsäcker's Long-Run Indifference Curves." *Journal of Economic Theory* 13:2 (October): 319–328.

Fieller, E. C. 1932. "The Distribution of the Index in a Normal Bivariate Population." *Biometrika* 24:3/4 (November): 428–440.

Fisher, Franklin M. 1987. "Household Equivalence Scales and Interpersonal Comparisons." *Review of Economic Studies* 54:3 (July): 519–524.

Fletcher, R. 1972. "Fortran Subroutines for Minimization by Quasi-Newton Methods," Theoretical Physics Division, Atomic Energy Research Establishment, Harwell, England.

Fourgeaud, Claude, and Andre Nataf. 1959. "Consommation en Prix et Revenu Réels et Théorie des Choix." *Econometrica* 27:3 (July): 329–354.

Friedman, Milton. 1962. *Price Theory: A Provisional Text.* Chicago: Aldine.

Gaertner, Wulf. 1974. "A Dynamic Model of Interdependent Consumer Behavior." *Zeitschrift für Nationalökonomie* 34:3–4: 327–344.

Galbraith, John Kenneth. 1958. *The Affluent Society.* Cambridge: Houghton Mifflin.

Galbraith, John Kenneth. 1970. "Economics as a System of Belief." *American Economic Review* 60:2 (May): 469–478.

Gillingham, Robert, and William S. Reece. 1979. "A New Approach to Quality of Life Measurement." *Urban Studies* 16:3 (October): 329–332.

Gillingham, Robert, and William S. Reece. 1980. "Analytical Problems in the Measurement of the Quality of Life." *Social Indicators Research* 7:1–4 (January): 91–101.

Gintis, Herbert. 1974. "Welfare Criteria with Endogenous Preferences: The Economics of Education." *International Economic Review* 15:1 (June): 415–430.

Goldberger, Arthur S. 1969. "Directly Additive Utility and Constant Marginal Budget Shares." *Review of Economic Studies* 36:2 (April): 251–254.

Goldberger, Arthur S., and Theodore Gamaletsos. 1970. "A Cross-Country Comparison of Consumer Expenditure Patterns." *European Economic Review* 1:3 (Spring): 357–396.

Goldman, Steven M., and Hirofumi Uzawa. 1964. "A Note on Separability in Demand Analysis." *Econometrica* 32:3 (July): 387–398.

Gorman, W. M. 1959. "Separable Utility and Aggregation." *Econometrica* 27:3 (July): 469–481.

Gorman, W. M. 1961. "On a Class of Preference Fields." *Metroeconomica* 13:2 (August): 53–56.

Gorman, W. M. 1967 "Tastes, Habits, and Choices." *International Economic Review* 8:2 (June): 218–222.

Gorman, W. M. 1968. "The Structure of Utility Functions." *Review of Economic Studies* 35:4 (October): 367–390.

Gorman, W. M. 1971a. "Apologia for a Lemma." *Review of Economic Studies* 38:1 (January): 114.

Gorman, W. M. 1971b. "Clontarf Revisited." *Review of Economic Studies* 38:1 (January): 116.

Gorman, W. M. 1976. "Tricks with Utility Functions." In *Essays in Economic Analysis: Proceedings of the 1975 AUTE Conference, Sheffield*, edited by M. J. Artis and A. R. Nobay, pp. 211–243. Cambridge: Cambridge University Press.

Gorman, W. M. 1981. "Some Engel Curves." In *Essays in the Theory and Measurement of Consumer Behaviour in Honour of Sir Richard Stone*, edited by Angus Deaton, pp. 7–29. Cambridge: Cambridge University Press.

Gronau, Reuben. 1988. "Consumption Technology and the Intrafamily Distribution of Resources: Adult Equivalence Scales Reexamined." *Journal of Political Economy* 96:6 (December): 1183–1205.

Halmos, Paul R. 1958. *Finite-Dimensional Vector Spaces*, 2nd ed. Princeton: van Nostrand.

Hammond, Peter J. 1976. "Endogeneous Tastes and Stable Long-Run Choice." *Journal of Economic Theory* 13:2 (October): 329–340.

Harsanyi, John C. 1953. "Cardinal Utility in Welfare Economics and in the Theory of Risk Taking." *Journal of Political Economy* 61:5 (October): 434–435.

Harsanyi, John C. 1955. "Cardinal Welfare, Individualistic Ethics, and Interpersonal Comparisons of Utility." *Journal of Political Economy* 63:4 (August): 309–321.

Harvey, A. C., and I. D. McAvinchey. 1978. "The Small Sample Efficiency of Two-step Estimators in Regression Models with Autoregressive Disturbance." Discussion Paper No 78–10, Department of Economics, University of British Columbia, (April).

Hayakawa, Hiroaki, and Yiannis Venieris, 1977. "Consumer Interdependence via Reference Groups." *Journal of Political Economy* 85:3 (June): 599–615.

Hicks, John R. 1969. "Direct and Indirect Additivity." *Econometrica* 37:2 (April): 353–354.

Houthakker, Hendrik S. 1960. "Additive Preferences." *Econometrica* 28:2 (April): 244–257.

Houthakker, Hendrik S. 1965. "A Note on Self-Dual Preferences." *Econometrica* 33:4 (October): 797–801.

Houthakker, Hendrik S., and John Haldi. 1960. "Household Investment in Automobiles." In *Consumption and Savings, Vol. I*, edited by Irwin Friend and Robert C. Jones, pp. 175–221. Philadelphia: University of Pennsylvania Press.

Houthakker, Hendrik S., and Lester D. Taylor. 1966. *Consumer Demand in the United States: Analyses and Projections*. Cambridge: Harvard University Press (2nd ed., 1970).

Howe, Howard. 1975. "Development of the Extended Linear Expenditure System from Simple Saving Assumptions." *European Economic Review* 6:3 (July): 304–310.

Howe, Howard, Robert A. Pollak, and Terence J. Wales. 1979. "Theory and Time Series Estimation of the Quadratic Ependiture System." *Econometrica* 47:5 (September): 1231–1247.

Hurwicz, Leonid, and Hirofumi Uzawa. 1971. "On the Integrability of Demand Functions." In *Preferences, Utility and Demand: A Minnesota Symposium*, edited by John S. Chipman, Leonid Hurwicz, Marcel K. Richter, and Hugo F. Sonnenschein, pp. 114–148. New York: Harcourt, Brace, Jovanovich.

Jorgenson, Dale W. 1990. "Aggregate Consumer Behavior and the Measurement of Social Welfare." *Econometrica* 58:5 (September): 1007–1040.

Jorgenson, Dale W., and Lawrence J. Lau. 1979. "The Integrability of Consumer Demand Functions." *European Economic Review*, 12:2 (April): 115–147.

Jorgenson, Dale W., Lawrence J. Lau, and Thomas M. Stoker. 1980. "Welfare Comparison under Exact Aggregation." *American Economic Review* 70:2 (May): 268–272.

Jorgenson, Dale W., and Daniel T. Slesnick. 1983. "Individual and Social Cost of

Living Indexes." In *Price Level Measurement,* edited by W. Erwin Diewert and Claude Montmarquette, pp. 241–323. Ottawa: Statistics Canada.

Klein, Lawrence R. 1962. *An Introduction to Econometrics.* Englewood Cliffs: Prentice Hall.

Klein, Lawrence R., and Herman Rubin. 1947–1948. "A Constant Utility Index of the Cost of Living." *Review of Economic Studies* 15:2: 84–87.

Klijn, Nico. 1977. "Expenditure, Savings, and Habit Formation: A Comment." *International Economic Review* 18:3 (October): 771–778.

Kravis, Irving B., Alan Heston, and Robert Summers. 1978. *International Comparisons of Real Product and Purchasing Power.* Baltimore: Johns Hopkins University Press.

Krelle, Wilhelm. 1973. "Dynamics of the Utility Function." In *Carl Menger and the Austrian School of Economics,* edited by John R. Hicks and Warren Weber, pp. 92–128. New York: Oxford University Press.

Lau, Lawrence J. 1977. "Existence Conditions for Aggregate Demand Functions." Technical Report No. 248, Institute for Mathematical Studies in the Social Sciences, Stanford University, Stanford.

Lau, Lawrence J. 1986. "Functional Forms in Econometric Model Building." In *Handbook of Econometrics, Vol. III,* edited by Zvi Griliches and Michael D. Intriligator, pp. 1515–1566. Amsterdam: Elsevier Science Publishers.

Lau, Lawrence J., Wuu-Long Lin, and Pan A. Yotopoulos. 1978. "The Linear Logarithmic Expenditure System: An Application to Consumption-Leisure Choice." *Econometrica* 46:4 (July): 843–868.

Lau, Lawrence J., and Bridger M. Mitchell. 1971. "A Linear Logarithmic Expenditure System: An Application to U.S. Data." *Econometrica* 39:4 (July): 87–88.

Lazear, Edward P., and Robert T. Michael. 1988. *Allocation of Income within the Household.* Chicago: University of Chicago Press.

Leibenstein, Harvey. 1950. "Bandwagon, Snob, and Veblen Effects in the Theory of Consumer Demand." *Quarterly Journal of Economics* 64:2 (May): 183–207.

Leibenstein, Harvey. 1975. "The Economic Theory of Fertility Decline." *Quarterly Journal of Economics* 89:1 (February): 1–31.

Leibenstein, Harvey. 1976. The Problem of Characterizing Aspirations." *Population and Development Review* 2:3/4 (September/December): 427–431.

Leontief, Wassily. 1947a. "A Note on the Interrelation of Subsets of Independent Variables of a Continuous Function with Continuous First Derivatives." *Bulletin of the American Mathematical Society* 53:4 (April): 343–350.

Leontief, Wassily. 1947b. "Introduction to a Theory of the Internal Structure of Functional Relationships." *Econometrica* 15:4 (October): 361–373.

Lewbel, Arthur. 1985. "A Unified Approach to Incorporating Demographic or Other Effects into Demand Systems." *Review of Economic Studies* 52:1 (January): 1–18.

Lewbel, Arthur. 1986. "Additive Separability and Equivalent Scales." *Econometrica* 54:1 (January): 219–222.

Lewbel, Arthur. 1987a. "Fractional Demand Systems." *Journal of Econometrics* 36:3 (November): 311–337.

Lewbel, Arthur. 1987b. "Characterizing Some Gorman Engel Curves." *Econometrica* 55:6 (November): 1451–1459.

Lewbel, Arthur. 1989. "Household Equivalence Scales and Welfare Comparisons." *Journal of Public Economics* 39:3 (August): 377–391.

Lewbel, Arthur. 1990. "Full Rank Demand Systems." *International Economic Review* 31:2 (May): 289–300.

Lluch, Constantino. 1973. "The Extended Linear Expenditure System." *European Economic Review* 4:1 (April): 21–31.

Lluch, Constantino. 1974. "Expenditure, Savings and Habit Formation." *International Economic Review* 15:3 (October): 786–797.

Luenberger, David G. 1979. *Introduction to Dynamic Systems.* New York: Wiley.

Lundberg, Shelly. 1988. "Labor Supply of Husbands and Wives: A Simultaneous Equations Approach." *Review of Economics and Statistics* 70:2 (May): 224–235.

Maasoumi, Esfandiar. 1986. "The Measurement and Decomposition of Multi-Dimensional Inequality." *Econometrica* 54:4 (July): 991–997.

Maeshiro, Asatoshi. 1979. "On the Retention of the First Observations in Serial Correlation Adjustment of Regression Models." *International Economic Review* 20:1 (February): 259–265.

Maeshiro, Asatoshi. 1980. "New Evidence on the Small Sample Properties of Estimators of SUR Models with Auto-correlated Disturbances: Things Done Halfway May Not Be Done Right." *Journal of Econometrics* 12:2 (February): 162–176.

Magnus, Jan R. 1982. "Multivariate Error Components Analysis of Linear and Nonlinear Regression Models by Maximum Likelihood." *Journal of Econometrics* 19:2/3 (August): 239–286.

Manser, Marilyn E. 1976. "Elasticities of Demand for Food: An Analysis Using Non-Additive Utility Functions Allowing for Habit Formation." *Southern Economic Journal* 43:1 (July): 879–891.

Manser, Marilyn E., and Murray Brown. 1980. "Marriage and Household Decision-Making: A Bargaining Analysis." *International Economic Review* 21:1 (February): 31–44.

McCarthy, Michael D. 1974. "On the Stability of Dynamic Demand Functions." *International Economic Review* 15:1 (February): 256–259.

McElroy, Marjorie B., and Mary J. Horney. 1981. "Nash–Bargained Household Decisions: Toward a Generalization of the Theory of Demand." *International Economic Review* 22:2 (June): 333–349.

Michael, Robert T., and Gary S. Becker. 1973. "On the New Theory of Consumer Behavior." *Swedish Journal of Economics* 75:4 (December): 378–396.

Modigliani, Franco. 1949. "Fluctuations in the Saving-Income Ratio: A Problem in Economic Forecasting." In *Studies in Income and Wealth, Vol. 11*, edited by Simon Goldberg and Phyllis Deane, pp. 371–402, 427–431. New York: National Bureau of Economic Research.

Muellbauer, John. 1974. "Household Composition, Engel Curves and Welfare Comparisons Between Households." *European Economic Review* 5:2 (August): 103–122.

Muellbauer, John. 1975. "Aggregation, Income Distribution and Consumer Demand." *Review of Economic Studies* 42:4 (October): 525–543.

Muellbauer, John. 1977. "Testing the Barten Model of Household Composition Effects and the Cost of Children." *Economic Journal* 87:347 (September): 460–487.

Muellbauer, John. 1980. "The Estimation of the Prais–Houthakker Model of Equivalence Scales." *Econometrica* 48:1 (January): 153–176.

Nerlove, Marc. 1974. "Household and Economy: Toward a New Theory of Population and Economic Growth." *Journal of Political Economy* 83:2, Part II (February): 44–62.

OECD National Accounts 1961–1978, Vol. 2. 1980. Paris: OECD.

Park, R. E., and Bridger M. Mitchell. 1979. "Maximum Likelihood vs. Minimum Sum of Squares Estimation of the Autocorrelated Error Model." N-1325 (November). Rand Corporation, Santa Monica, CA.

Park, R. E., and Bridger M. Mitchell. 1980. "Estimating the Autocorrelated Error Model with Trended Data: Further Results." *Journal of Econometrics* 13:2 (June): 185–201.

Parks, Richard W., and Anton P. Barten. 1973. "A Cross-Country Comparison of the Effects of Prices, Income and Population Composition on Consumption Patterns." *Economic Journal* 83:331 (September): 834–852.

Pashardes, Panos. 1986. "Myopic and Forward Looking Behavior in a Dynamic Demand System." *International Economic Review* 27:2 (June): 387–389.

Peleg, Bezalel, and Menahem E. Yaari. 1973. "On the Existence of a Consistent Course of Action When Tastes Are Changing." *Review of Economic Studies* 40:3 (July): 391–401.

Pesaran, M. Hashem, and Angus S. Deaton. 1978. "Testing Non-nested Nonlinear Regression Models." *Econometrica* 46:3 (May): 677–694.

Peston, Maurice H. 1967. "Changing Utility Functions." In *Essays in Mathematical Economics in Honor of Oskar Morgenstern*, edited by Martin Shubik, pp. 233–236. Princeton: Princeton University Press.

Phlips, Louis. 1972. "A Dynamic Version of the Linear Expenditure Model." *Review of Economics and Statistics* 54:4 (November): 450–458.

Phlips, Louis. 1974. *Applied Consumption Analysis*. Amsterdam: North-Holland (revised and enlarged edition, 1983).

Poirier, Dale J. 1978. "The Effect of the First Observation in Regression Models with First Order Autoregressive Disturbances." *Applied Statistics* 27:1 (January): 67–68.

Pollak, Robert A. 1968. "Consistent Planning." *Review of Economic Studies* 35:2 (April): 201–208.

Pollak, Robert A. 1969. "Conditional Demand Functions and Consumption Theory." *Quarterly Journal of Economics* 83:1 (February): 60–78.

Pollak, Robert A. 1970. "Habit Formation and Dynamic Demand Functions." *Journal of Political Economy* 78:4 (July/August): 745–763.

Pollak, Robert A. 1971a. "Conditional Demand Functions and the Implications of Separable Utility." *Southern Economic Journal* 37:4 (April): 423–433.

Pollak, Robert A. 1971b. "Additive Utility Functions and Linear Engel Curves." *Review of Economic Studies* 38:4 (October): 401–414.

Pollak, Robert A. 1972. "Generalized Separability." *Econometrica* 40:3 (May): 431–453

Pollak, Robert A. 1974. "The Treatment of the Environment in the Cost-of-Living Index." Mimeograph. Reprinted in Pollak (1989).

Pollak, Robert A. 1975. "The Intertemporal Cost of Living Index." *Annals of Economic and Social Measurement* 4:1 (Winter): 179–195. Reprinted in Pollak (1989).

Pollak, Robert A. 1976a. "Interdependent Preferences." *American Economic Review* 66:3 (June): 309–320.

Pollak, Robert A. 1976b. "Habit Formation and Long-Run Utility Functions." *Journal of Economic Theory* 13:2 (October): 272–297.

Pollak, Robert A. 1977. "Price Dependent Preferences." *American Economic Review* 67:2 (March): 64–75.

Pollak, Robert A. 1978. "Endogenous Tastes in Demand and Welfare Analysis." *American Economic Review* 68:2 (May): 374–379.

Pollak, Robert A. 1981. "The Social Cost of Living Index." *Journal of Public Economics* 15:3 (June): 311–336. Reprinted in Pollak (1989).

Pollak, Robert A. 1985. "A Transaction Cost Approach to Families and Households." *Journal of Economic Literature* 23:2 (June): 581–608.

Pollak, Robert A. 1989. *The Theory of the Cost-of-Living Index.* New York: Oxford University Press.

Pollak, Robert A. 1990. "Distinguished Fellow: Houthakker's Contributions to Economics." *Journal of Economic Perspectives* 4:2 (Spring): 41–156.

Pollak, Robert A. 1991. "Welfare Comparisons and Situation Comparisons." *Journal of Econometrics* 50:1/2 (October/November): 31–48.

Pollak, Robert A., Robin C. Sickles, and Terence J. Wales. 1984. "The CES-translog: Specification and Estimation of a New Cost Function." *Review of Economics and Statistics* 66:4 (November): 602–607.

Pollak, Robert A., and Michael L. Wachter. 1975. "The Relevance of the Household Production Function and Its Implications for the Allocation of Time." *Journal of Political Economy* 83:2 (April): 255–277.

Pollak, Robert A., and Terence J. Wales. 1969. "Estimation of the Linear Expenditure System." *Econometrica* 37:4 (October): 611–628.

Pollak, Robert A., and Terence J. Wales. 1978. "Estimation of Complete Demand Systems from Household Budget Data: The Linear and Quadratic Expenditure Systems." *American Economic Review* 68:3 (June): 348–359.

Pollak, Robert A., and Terence J. Wales. 1979. "Welfare Comparisons and Equivalence Scales." *American Economic Review* 69:2 (May): 216–221. Reprinted in Pollak (1989).

Pollak, Robert A., and Terence J. Wales. 1980. "Comparison of the Quadratic Expenditure System and Translog Demand Systems with Alternative Specifications of Demographic Effects." *Econometrica* 48:3 (April): 595–612.

Pollak, Robert A., and Terence J. Wales. 1981. "Demographic Variables in Demand Analysis." *Econometrica* 49:6 (November): 1533–1551.

Pollak, Robert A., and Terence J. Wales. 1982. "Demographic Variables in Demand Analysis and Welfare Analysis." In *Economic Activity and Finance*, edited by Marshall E. Blume, Jean Crockett, and Paul Taubman, pp. 85–125. Cambridge: Ballinger.

Pollak, Robert A., and Terence J. Wales. 1987. "Pooling International Consumption Data." *Review of Economics and Statistics* 69:1 (February): 90–99.

Pollak, Robert A., and Terence J. Wales. 1991. "The Likelihood Dominance Criterion: A New Approach to Model Selection." *Journal of Econometrics* 47:2/3 (February/March): 227–242.

Pollak, Robert A., and Terence J. Wales. 1992. "Specification and Estimation of Dynamic Demand Systems." In *Aggregation, Consumption, and Trade:*

Essays in Honor of H. S. Houthakker, edited by Louis Phlips and Lester D. Taylor. Boston: Kluwer, 1992.

Prais, S. J., and Hendrik S. Houthakker. 1955. *The Analysis of Family Budgets.* Cambridge: Cambridge University Press.

Rothenberg, T. J., and C. T. Leenders. 1964. "Efficient Estimation of Simultaneous Equation Systems." *Econometrica* 32:1–2 (January): 57–76.

Samuelson, Paul A. 1947. *Foundations of Economic Analysis.* Cambridge: Harvard University Press.

Samuelson, Paul A. 1947–1948. "Some Implications of 'Linearity'." *Review of Economic Studies* 15:2: 88–90.

Samuelson, Paul A. 1956. "Social Indifference Curves." *Quarterly Journal of Economics* 70:1 (February): 1–22.

Samuelson, Paul A. 1969. "Corrected Formulation of Direct and Indirect Additivity." *Econometrica* 37:2 (April): 355–359.

Sen, Amartya. 1970. *Collective Choice and Social Welfare.* San Francisco: Holden-Day.

Sen, Amartya. 1973. *On Economic Inequality.* Oxford: Clarendon Press.

Sen, Amartya. 1976. "Poverty: An Ordinal Approach to Measurement." *Econometrica* 44:2 (March): 219-231.

Sen, Amartya. 1982. *Choice, Welfare and Measurement.* Cambridge: MIT Press.

Sen, Amartya. 1985. *Commodities and Capabilities.* Amsterdam: North-Holland.

Sen, Amartya. 1987. *The Standard of Living.* Cambridge: Cambridge University Press.

Spinnewyn, Frans. 1981. "Rational Habit Formation." *European Economic Review* 15:1 (January): 91–109.

Stigler, George J., and Gary S. Becker. 1977. "De Gustibus Non Est Disputandum." *American Economic Review* 67:2 (March): 76–90.

Stone, Richard. 1954. "Linear Expenditure Systems and Demand Analysis: An Application to the Pattern of British Demand." *Economic Journal* 64:255 (September): 511–527.

Stone, Richard. 1966a. "The Changing Pattern of Consumption." In *Mathematics in the Social Sciences and Other Essays,* edited by Richard Stone, pp. 190–203. Cambridge: MIT Press.

Stone, Richard. 1966b. "British Economic Balances in 1970: A Trial Run on Rocket." In *Mathematics in the Social Sciences and Other Essays,* edited by Richard Stone, pp. 249–282. Cambridge: MIT Press.

Strotz, Robert H. 1955–1956. "Myopia and Inconsistency in Dynamic Utility Maximization." *Review of Economic Studies* 23:3: 165–180.

Strotz, Robert H. 1957. "The Empirical Implications of a Utility Tree." *Econometrica* 25:2 (April): 269–280.

Strotz, Robert H. 1959. "The Utility Tree—A Correction and Further Appraisal." *Econometrica* 27:3 (July): 482–488.

Sydenstricker, Edgar, and Wilford I. King. 1921. "The Measurement of Relative Economic Status of Families." *Quarterly Publication of the American Statistical Association* 17:135 (September): 842–857.

Taylor, Lester D., and Daniel Weiserbs. 1972. "On the Estimation of Dynamic Demand Functions." *Review of Economics and Statistics* 54:4 (November): 459–465.

Theil, Henri. 1971. *Principles of Econometrics.* New York: Wiley.

U. K. Central Statistical Office. 1974. *Annual Abstract of Statistics*. London: HMSO.

U. K. Central Statistical Office. 1975. *National Income and Expenditure 1964–74*. London: HMSO.

U. K. Department of Employment and Productivity. Various years. *Family Expenditure Survey*. London: HMSO.

United Nations Demographic Yearbook. 1979. New York: United Nations.

van Daal, Jan, and Arnold H. Q. M. Merkies. 1989. "A Note on the Quadratic Expenditure Model." *Econometrica* 57:6 (November): 1439–1443.

Varian, Hal R. 1982. "The Nonparametric Approach to Demand Analysis." *Econometrica* 50:4 (July): 945–973.

Vind, Karl. 1971a. "Note on 'The Structure of Utility Functions'." *Review of Economic Studies* 38:1 (January): 113.

Vind, Karl. 1971b. "Comment." *Review of Economic Studies* 38:1 (January): 115.

von Weizsäcker, Carl Christian. 1971. "Notes on Endogenous Change of Tastes." *Journal of Economic Theory* 3:4 (December): 345–372.

Wales, Terence J. 1971. "A Generalized Linear Expenditure Model of the Demand for Non-Durable Goods in Canada." *Canadian Journal of Economics* 4:4 (November): 471–484.

Wales, Terence J. 1984. "A Note on Likelihood Ratio Tests of Functional Form and Structural Change in Demand Systems." *Economic Letters* 14:2–3: 213–220.

Wales, Terence J., and Alan D. Woodland. 1976. "Estimation of Household Utility Functions and Labor Supply Response." *International Economic Review* 17:2 (June): 397–410.

Yatchew, Adonis J. 1985. "Labor Supply in the Presence of Taxes: An Alternative Specification." *Review of Economics and Statistics* 67:1 (February): 27–33.

Yatchew, Adonis J. 1986. "Multivariate Distributions Involving Ratios of Normal Variables." *Communications in Statistics, Theory, and Methods* 15:6: 1905–1926.

Subject Index

Author Index